DREISER AND VEBLEN
SABOTEURS OF THE STATUS QUO

DREISER
SABOTEURS OF

AND VEBLEN
THE STATUS QUO

Clare Virginia Eby

UNIVERSITY OF MISSOURI PRESS
COLUMBIA AND LONDON

Copyright © 1998 by
The Curators of the University of Missouri
University of Missouri Press, Columbia, Missouri 65201
Printed and bound in the United States of America
All rights reserved
5 4 3 2 1   02 01 00 99 98

Library of Congress Cataloging-in-Publication Data

Eby, Clare Virginia.
   Dreiser and Veblen, saboteurs of the status quo / Clare Virginia Eby.
     p. cm.
   Includes bibliographical references and index.
   ISBN 0-8262-1193-3 (alk. paper)
   1. Dreiser, Theodore, 1871–1945—Political and social views.
2. Literature and society—United States—History—20th century.
3. Veblen, Thorstein, 1857–1929—Influence.   4. Social problems—
United States—History.   5. Social problems in literature.
I. Title.
PS3507.R55Z625   1998
813'.52—dc21
                                                98-44703
                                                    CIP

∞™ This paper meets the requirements of the
American National Standard for Permanence of Paper
for Printed Library Materials, Z39.48, 1984.

Text Design: Stephanie Foley
Jacket Design: Susan Ferber
Typesetter: BOOKCOMP
Printer and Binder: Thomson Shore
Typefaces: Fenice and Garamond

For permissions see p. 227.

To my parents

CONTENTS

PREFACE   ix
ACKNOWLEDGMENTS   xiii
ABBREVIATIONS OF WORKS FREQUENTLY CITED   xvii

Introduction   1

1. The Rhetoric of Confrontation   20

2. Business as (Un)usual
   The Immaterial Economy in *The Trilogy of Desire* 65

3. The Psychology of Desire
   Pecuniary Emulation and Invidious Comparison in
   Sister Carrie and *An American Tragedy*   107

4. Gender and Cultural Criticism
   The Instinct of Workmanship in *Jennie Gerhardt* 148

BIBLIOGRAPHY   191
INDEX   219

PREFACE

"With Veblen perhaps the whole character of American social criticism shifted. The figure of the last-generation American faded and the figure of the first-generation American—the Norwegian immigrant's son, the New York Jew teaching English literature in a midwestern university, the southerner come north to crash New York—was installed as the genuine, if no longer 100 per cent American, critic."
—*C. Wright Mills*

"The people who were most American by birth . . . gave themselves a literature which had the least to say about the real phenomena of American life, most particularly the accelerated rate, the awful rate, of growth and anomaly through all of society. That sort of literature and that kind of attempt to explain America was left to the sons of immigrants. . . . [Dreiser] came closer to understanding the social machine than any American writer who ever lived, but he paid an unendurable price—he was forced to alienate himself from manner in order to learn the vast amount he learned. . . . Dreiser was in a huge hurry, he had to learn everything . . . so there is nothing of manner in his work."
—*Norman Mailer*

C. Wright Mills and Norman Mailer both point to the social upheavals confronting those who came of age during "the incorporation of America," and to the invention of new discursive modes to address those transformations. Mills describes the unique style of cultural criticism pioneered by Thorstein Veblen (1857-1929), and Mailer, the massive but ill-digested learning of Theodore Dreiser (1871–1945), as reactions to and representations of social change. Claiming that it took children of immigrants who were, in Mills's words, less than "100 per cent American" to come to terms with what Mailer calls "the social machine" may sound problematic today. But there is nothing nativist or parochial in Cornel West's charge that the booming field of cultural studies in the United States needs to rediscover its history on native grounds. West mentions Veblen as an American thinker whom contemporary cultural analysts need to examine, a claim I second, and extend also to Dreiser.[1]

The works of Dreiser and Veblen make up a neglected chapter in the history of United States cultural criticism. Their central subjects (such as the myriad effects of consumer capitalism and the invidious status system) still preoccupy cultural critics, and with good reason. Veblen and Dreiser also pioneered strategies for positioning themselves as confrontational intellectuals (such as by attacking foundationalism and claims of epistemological certainty) that continue to inform the practice of many cultural critics. Thus, in both subject matter and rhetorical strategy, Dreiser's and Veblen's writings provide prototypes for the work that many United States scholars want to do now, work which often turns to European or postmodern theory for inspiration. In making this claim about the usefulness of Dreiser and Veblen for current intellectual work, my argument parallels recent rehabilitations of American thinkers, such as Richard Rorty's contention that John Dewey and William James are "waiting at the end of the road which . . . Foucault and Deleuze are currently traveling," and Frank Lentricchia's explanation of why James is "recapturing the imagination of the American literary humanist from its fascination with the French intellectual scene."[2]

This is not to say that Veblen and Dreiser are closet postmodernists.[3] While Veblen's majestic skepticism toward all claims of epistemological

---

1. The phrase is Alan Trachtenberg's; see *The Incorporation of America: Culture and Society in the Gilded Age;* C. Wright Mills, introduction to the Mentor edition of *The Theory of the Leisure Class,* xi; Norman Mailer, *Cannibals and Christians,* 97; Cornel West, "The Postmodern Crisis of the Black Intellectuals," 693.

2. Richard Rorty, *Consequences of Pragmatism,* xviii; Frank Lentricchia, *Ariel and the Police: Michel Foucault, William James, Wallace Stevens,* 104.

3. In *Social Theory and Social Structure,* Robert K. Merton cautioned many years ago that the search for "anticipations" of ideas in earlier periods forms an "occupational disease" of certain types of historical investigation (14).

certainty may impress even the most radical of today's cultural critics, his rhetoric of disinterestedness will not; and while Dreiser's understanding of the social construction and policing of sexuality predates Foucault's, his heterosexual biases and essentializing tendencies limit the utility of his ideas for current discussions of sexuality. My interests are primarily historical and contextual, but they are informed by the belief that the cultural criticism of Dreiser and Veblen remains usable today and continues to be of historical importance even when needing revision.

Dreiser and Veblen deserve our attention also because the striking parallels between their works illustrate how literature and social science can come together on the ground of cultural criticism. Dreiser's canon might well be described as sociological belles lettres, while Veblen's corpus makes up one of the most rhetorically loaded and stylistically distinctive instances of social theory. They have various and distinctive ways of using cultural criticism to reconnoiter the lines believed to separate literature from social science, and so by placing their works into dialogue, I want to contribute to the recent attempts of literary analysts and social scientists to talk to one another.

The explosion of interest in interdisciplinary scholarship needs to be accompanied by historical investigation of earlier thinkers—such as Veblen and Dreiser—who crossed disciplinary lines. The vast scope of their writings has been obscured by the narrow frameworks through which they have been interpreted and by the related tendency to privilege certain of their works to the virtual exclusion of all others. Most approaches to Dreiser assume some definition of the "realist" or "naturalist" novel as the most salient reference point, while studies of Veblen generally begin by situating him within a particular social scientific discipline, typically economics or sociology. (The notoriety of a single book, *The Theory of the Leisure Class,* has further obscured the range of Veblen's ideas, while Dreiser's views are most often assessed through *Sister Carrie* and *An American Tragedy.*) Drawing extensively from the works of Veblen and Dreiser, I use cultural criticism as a unifying concept that shows how Veblen fuses sociology, economics, history, psychology, anthropology, political science, satire, and philosophy; and how Dreiser conjoins fiction, travelogue, literary manifesto, occasional essay, autobiography, biography, and philosophy. In their interdisciplinary scope, Dreiser and Veblen resemble the "all-purpose intellectuals" that Richard Rorty calls for, thinkers "ready to offer a view on pretty much anything." And the work of the all-purpose intellectual, Rorty notes, looks a great deal like cultural criticism.[4]

---

4. Rorty, *Consequences,* xxxix, xl.

ACKNOWLEDGMENTS

I first began exploring Dreiser and Veblen while writing my dissertation at the University of Michigan on businessmen as characters in turn-of-the-century American fiction. I was lucky to have had June Howard as my advisor; her continual encouragement to take a more interdisciplinary approach has shaped my thinking for good. Although neither she nor the other members of my committee read any of this manuscript, I learned from them all, especially David Lewis and the late James Gindin.

Several years after moving to Connecticut, I was lucky enough to become involved with the newly formed International Thorstein Veblen Association. Eternal thanks go to Clint Sanders for telling me about the association. The warm welcome I received from this lively group convinced me that it was possible for academics to have meaningful discussions across disciplinary lines. Of my Veblen "clansmen," Arthur Vidich and Rick Tilman have been particularly helpful, gracious, and supportive of my work. Stephen Edgell, Jack Diggins, Howard Horwitz, Beverly Haviland, and others who participated in recent conferences have asked tough questions that forced me to think more clearly about Veblen. Larry Van Sickle and Arthur Vidich have each read parts of this manuscript and helped me understand the questions that sociologists want answered.

I have also profited in numerous ways from associations with individuals in the Dreiser Society. I am grateful to Phil Gerber for organizing the "Working Girls" conference to commemorate the ninetieth anniversary of *Sister Carrie,* at which I first met many of the people who presently write about Dreiser. My thanks to Jim West for useful comments on a short version of the chapter on *Jennie Gerhardt* and for consistently sound advice over the years. Jim Hutchisson also read an earlier version of chapter 2. Larry Hussman continues to inspire me by demonstrating that the smartest academics retain a sense of humor. (Working with Dreiser and Veblen, one needs that.)

As readers for an essay that first appeared in *American Quarterly* and is now part of chapter 1, David Hollinger and David Noble offered good counsel, as did anonymous readers for *Studies in American Fiction*, which published an early version of chapter 3. Two readers for the *Journal of Economic Issues* sent me back to the library to rethink what I had said about Veblen in an article unrelated to this project, and their suggestions helped me bring this book to a close.

Many people at the University of Connecticut have sustained my interest in this project either directly or indirectly. Veronica Makowsky generously read the entire manuscript in a relatively late form and showed me how to make it better. Eleni Coundouriotis took time out from her own last-minute book revisions to read one of the chapters, allaying my anxieties while making some wise suggestions. Jerry Phillips is one of few people here with whom I can actually talk about Veblen, and I'm grateful to have him as a colleague. Tom and Pegi Shea, Arnold Orza, Mary Gallucci, Robin Simmons, Pam Stockamore, Lynn Bloom, and Tom Cooke have all been supportive friends and colleagues over the years. I am also most grateful to the many undergraduate and graduate students who have read Dreiser and Veblen along with me; their interest, resistance, and skepticism helped to keep me certain that these writers deserved attention and that what I had to say about them wasn't self-evident.

Two colleagues have played starring roles in keeping me going as I wrote this book. In the many times Lenny Cassuto has gone over chunks of the manuscript, he has proven himself as talented at microediting as at reconceptualizing ideas. Equally skilled at knowing what matters in our profession, Lenny's greatest genius of all may be in friendship. That he and Debra Osofsky have hosted me during various research and conference trips only increases a debt I happily acknowledge. Debts I can never possibly repay are drawn on the account of Tom Riggio, my mentor since I came to Connecticut—a mentor, I might add, who could not possibly be better suited to my needs. Tom is an exemplary colleague whose advice on writing, teaching, publishing, and surviving academia has strengthened me immeasurably.

As the above paragraph reminds me, personal associations have been essential in sustaining this long-term project. Family members and friends have continued to believe in it (and me) even when I faltered: first and foremost, my sister Lillian Eby and grandmother Ala K. McGuire, and also Craig Pascoe, Jack Aldridge, Nora Arato, June and Dan Lo Presti, Georgia Lo Presti, Harriet Feldlaufer, David Reuman, John and Ginger Nastasi (the latter my PR agent on the East Coast), and many others. My husband, John Lo Presti, who looked after me on a daily basis during the eight years I've worked on this book, rightfully deserves a dedication, but he will have to wait for my second book (which hopefully will not take so long to write). My deepest debt is to my parents, Patricia Aldridge and Cecil Eby, to whom I dedicate this book. They taught me to value intellectual inquiry and to understand that anything of value takes a long time to accomplish, and perhaps most important, they raised me in such a way that it never dawned on me to think I couldn't write a book. I am well aware that few daughters are so fortunate.

Although these pages have a lot to say about what's wrong with institutions, I must admit that several have assisted me. The University of Connecticut Research Foundation gave me a Junior Faculty Grant that launched this project and has supported it with several travel grants. A sabbatical allowed me the time to start pulling the many strands of this book together. A subsequent University of Connecticut Provost's Fellowship relieved me from teaching for a second term in which I began archival research on a new project, and that research helped this book as well. The staff of the Trecker Library of the University of Connecticut at Hartford has cheerfully and expeditiously fulfilled hundreds of requests for books and articles; Jan Lambert, in particular (and previously Lois Fletcher), has spoiled me rotten and enabled my research in countless ways. My thanks go also to the professionals who assisted my access to documents at the Rare Book and Manuscript Library at Columbia University, the Minnesota Historical Society, the State Historical Society of Wisconsin, the Van Pelt-Dietrich Library at the University of Pennsylvania, and the Manuscript Division of the Library of Congress. A special thanks goes to Eric Hilleman, the knowledgeable Carleton College Archivist, and also to the gentleman I serendipitously met at the Historical Society of Wisconsin who suggested I look in the Rasmus B. Anderson papers for material about Veblen.

The University of Missouri Press has done an exemplary job of guiding me through the publication process and producing this book expeditiously. I am particularly grateful to Clair Willcox, for his initial interest in the manuscript, and Sara Davis, for her careful editing of it.

Needless to say, the limitations of this work, of which I am well aware, are exclusively my own.

ABBREVIATIONS OF WORKS FREQUENTLY CITED

Dreiser Titles

Theodore Dreiser, *An Amateur Laborer*	*AL*
American Diaries 1902-1926	*AD*
An American Tragedy	*AAT*
A Book about Myself (*Newspaper Days*, 1922 edition)	*BM*
Dawn	*D*
Dreiser Looks at Russia	*DLR*
The Financier (1912 edition)	1912 *F*
The Financier (1927 edition)	*F*
The "Genius"	*G*
Hey Rub-a-Dub-Dub	*HRDD*
A Hoosier Holiday	*HH*
Jennie Gerhardt	*JG*
Newspaper Days (1991 Pennsylvania edition)	*ND*
Notes on Life	*NL*
Sister Carrie (1900 Doubleday edition)	*SC*
Sister Carrie (1981 Pennsylvania Edition)	Penn *SC*

The Stoic   S

The Titan   T

Theodore Dreiser: A Selection of Uncollected Prose   SUP

Veblen Titles

Thorstein Veblen, *Absentee Ownership and Business Enterprise in Recent Times*	AO
The Engineers and the Price System	EPS
Essays in Our Changing Order	ECO
The Higher Learning in America	HLA
Imperial Germany and the Industrial Revolution	IG
An Inquiry into the Nature of Peace and the Terms of its Perpetuation	NOP
The Instinct of Workmanship and the State of the Industrial Arts	IOW
The Place of Science in Modern Civilization	POS
The Theory of Business Enterprise	TBE
The Theory of the Leisure Class	TLC
The Vested Interests and the Common Man	VI

DREISER AND VEBLEN
SABOTEURS OF THE STATUS QUO

Introduction

"Any anthology of American prose in the future and any history of American literature will ignore Veblen at its peril."
—*Max Lerner, "Veblen's World"*

"The thing [Dreiser's *Jennie Gerhardt*] is not a mere story, not a novel in the ordinary American meaning of the word, but a criticism and an interpretation of life."
—*H. L. Mencken, "A Novel of the First Rank"*

It is difficult to imagine a more Veblenian vantage point than Theodore Dreiser's view of Sherry's Restaurant in *Sister Carrie* (1900). Dreiser's description of the stylish New York eatery as a place for "showy, wasteful, and unwholesome gastronomy" (*SC*, 297) could be a page torn from Thorstein Veblen's analysis of conspicuous consumption and reputable waste in *The Theory of the Leisure Class*, published one year earlier. This sort of congruity led Richard Hofstadter to call Veblen "some sort of analogue of Dreiser," an insight that has often been echoed but never pursued systematically.[1] Dreiser and Veblen share many rhetorical strategies, numerous subjects, and most significant, a

---

1. Richard Hofstadter quoted in David Riesman, *Thorstein Veblen: A Critical Interpretation,* 7. David Noble's "Dreiser and Veblen and the Literature of Cultural Change" offers the most extensive treatment of the two. Two dissertations, Catherine Ann Caraher's "Thorstein Veblen and the American Novel" and Robert Morton McIlvaine's "Thorstein Veblen and American Naturalism," include chapters on Dreiser. Briefer comparison of the two figures include Vernon L. Parrington, *Main Currents in American Thought,* vol. 3: *The Beginnings of Critical Realism in America: 1860-1920;* Alfred

way of viewing United States society that helped to define twentieth-century cultural criticism. This study illustrates these analogies by way of reciprocal interpretation: I use Dreiser to provide concrete illustrations of Veblen, whose convoluted prose can make his ideas difficult to grasp, and Veblen's theories to disclose the consistency and acuity of Dreiser's understanding of social scientific issues. In the process of providing a new way of seeing Dreiser and Veblen, both individually and together, I will also illustrate some of the tensions within the culture that they describe (many of which persist), as well as tensions within the roles they develop for themselves as cultural critics.

Two distinctive aspects of Veblen's critical social theory, his institutionalism and his belief in the stability of the class system in the United States, can serve to introduce the fundamental analogy with Dreiser and define the sort of cultural criticism that both practice. Veblen's intellectual legacy is presently most visible in the institutional school of economics, which is the leading heterodox but non-Marxist branch of the dismal science in the United States.[2] Yet even if Veblen were assumed to be primarily an economist (a confinement I should not care to enforce), the central point of his institutional approach is that economic activities cannot be isolated from other cultural practices.

Kazin, *On Native Grounds: An Interpretation of Modern American Prose;* Riesman, *Thorstein Veblen;* Philip Fisher, *Hard Facts: Setting and Form in the American Novel;* Susan Mizruchi, *The Power of Historical Knowledge: Narrating the Past in Hawthorne, James, and Dreiser.*

On Veblen and other contemporaneous literary figures, see Ross Posnock, "Henry James, Veblen, and Adorno: The Crisis of the Modern Self"; Beverly Haviland, "Waste Makes Taste: Thorstein Veblen, Henry James, and the Sense of the Past"; Daniel Lance Bratton, "Conspicuous Consumption and Conspicuous Leisure in the Novels of Edith Wharton"; Ruth Bernard Yeazell, "The Conspicuous Wasting of Lily Bart"; Howard Horwitz, *By the Law of Nature: Form and Value in Nineteenth Century America;* Mark Schorer, *Sinclair Lewis: An American Life;* my "*Babbitt* as Veblenian Critique of Manliness"; John Diggins, "Dos Passos and Veblen's Villains"; and Thomas Galt Peyser, "Reproducing Utopia: Charlotte Perkins Gilman and *Herland.*"

2. The phrase *institutionalist* as a designation of economic thought derives from Veblen; he subtitles *Leisure Class* "An Economic Study of Institutions." Also self-described as evolutionary economists, to distinguish their perspective from the equilibrium-oriented approaches, institutionalists follow Veblen in taking up a huge variety of issues, from historical comparisons among thinkers to aesthetics to analyses of consumption. Many commentators distinguish between "old" institutionalism, associated with Veblen, John Commons, and Wesley Mitchell, and neoinstitutionalism. For a concise definition of "old" institutionalism, see Geoffrey Hodgson, *Economics and Evolution: Bringing Life Back into Economics,* 301, n. 1. Also see Geoffrey Hodgson's "Institutional Economics: Surveying the 'Old' and the 'New.'"

Thus in the most famous course he taught at the University of Chicago, "Economic Factors in Civilization," Veblen discussed not only currency and business organization, but also religion, art, literature, race theory, the family, philosophy, psychology, and Viking raids.[3] Recent interdisciplinary scholarship appropriately situates Veblen's writings within a range of cultural practices, including technocracy and social engineering, the "double discourse" of the body and the machine, and changing notions of personal, ethical, and economic values.[4]

Dreiser's thinking is as interdisciplinary in method and institutional in orientation as Veblen's. The obsession of Dreiser's characters with money, social class, and consumer goods, which is one of his hallmarks, continues to generate new interpretations precisely because economic activity points in so many directions in his novels, as it does in Veblen's works. Dreiser understands not only the multiple effects of economic status on individual characters' lives (it influences everything from erotic desire to familial relations), but also how class positions determine the broader trends of United States life. As the latest generation of Dreiser scholars has made clear, it is not that he reduces everything to cash value but that the economic nexus reveals so much about personality formation, social organization, aesthetics, philosophy, and narrative structure, among other things.[5]

I want to draw attention to how the institutional approach shared by Dreiser and Veblen has strong family ties with cultural criticism. Their interdisciplinary perspective provides an ever-shifting vantage point from which to criticize specific cultural practices. The institutional

---

3. Joseph Dorfman, *Thorstein Veblen and His America*, 239.

4. See Martha Banta, *Taylored Lives: Narrative Productions in the Age of Taylor, Veblen and Ford;* John M. Jordan, *Machine-Age Ideology: Social Engineering and American Liberalism, 1911-1939;* Mark Seltzer, *Bodies and Machines;* Horwitz, *By the Law of Nature.*

5. In "The Country of the Blue," Eric Sundquist describes "the age of realism [as] the age of the *romance of money*" (19)—a designation particularly apt for Dreiser. On Dreiser and economics see, besides the studies cited in note 1: Amy Kaplan, *The Social Construction of American Realism;* Rachel Bowlby, *Just Looking: Consumer Culture in Dreiser, Gissing, and Zola;* June Howard, *Form and History in American Literary Naturalism;* Thomas Strychacz, *Modernism, Mass Culture, and Professionalism;* Walter Benn Michaels, *The Gold Standard and the Logic of Naturalism: American Literature at the Turn of the Century.* On literary realism and economics, see also Trachtenberg, *Incorporation of America;* Daniel Borus, *Writing Realism: Howells, James, and Norris in the Mass Market;* Christopher Wilson, *The Labor of Words: Literary Professionalism in the Progressive Era;* Michael Anesko, *"Friction with the Market": Henry James and the Profession of Authorship.*

viewpoint also lets an analyst distinguish between an individual actor's *motive* for doing something and the *social function* of her behavior, and in doing so breaks decisively with an earlier and more individualistic model of cultural criticism as practiced, for instance, by Ralph Waldo Emerson and William James. No longer are institutions to be seen as the lengthened shadows of individuals, as Emerson famously declared. Rather, institutionalism posits a reciprocal interaction between individuals and social structures: society cannot exist without individuals, nor can individuals exist prior to (or independent of) social structures. For example, my motive for buying faux leopard print clothing may be to appear stylishly retro while sending out signals that I have a wild side beyond my professorial role, but the social functions of my purchases include perpetuating the notion of women as sexual prey as well as greasing the wheels of the lucrative fashion machine. Thus, from the standpoint of social function, an expression of "individuality" in fact reveals the construction of personality by sex/gender codes and by capitalism. Using institutional analysis to identify such tensions can facilitate cultural criticism. As Robert Merton observed many years ago, the distinction between individual intent and social function frequently has an unmasking effect and generates conclusions that challenge prevailing views.[6] Or, to use Veblen's words, because institutions "are products of the past process, are adapted to past circumstances, and are therefore never in full accord with the requirements of the present," they tend inevitably to become "imbecile" (*TLC*, 191; *IOW*, 25). Dreiser, who uses the related term *conventions* more often than *institutions*, also sees them as decrepit and disabling. Dreiser and Veblen's cultural criticism reveals how institutions function to circumscribe individual actions.[7]

A second defining factor of Veblen's thinking that links it with Dreiser's perspective (while distinguishing both from Marx's more

---

6. Merton, *Social Theory*, 124-25. Veblen's disclaimer about his use of the word *waste* provides a good example. He pretends to apologize for his acerbic comments on the latent social function of conspicuous consumption—such as "In order to be reputable it [an expenditure] must be wasteful"—by noting, "The use of the term 'waste' is in one respect an unfortunate one. As used in the speech of everyday life the word carries an undertone of deprecation" (*TLC*, 96, 97).

7. My thinking here is indebted to Merton's distinction between manifest and latent function in *Social Theory*, chap. 3. For a recent attempt to rehabilitate institutionalism, drawing largely on the work of Emile Durkheim and Ludwik Fleck, see Mary Douglas, *How Institutions Think*.

celebrated analysis) is the assumption that the class structure will remain in place in the United States. Despite inequalities in the distribution of wealth as dramatic today as at the turn into the twentieth century, most Americans still believe the country provides equal opportunities for all citizens to succeed. Veblen provides a compelling explanation of this American enthusiasm for the invidious class structure through his analyses of emulation, consumption, and institutional lag. As long as we strive to consume more than our neighbors do, we keep the capitalist engine moving and the class structure in place.

Dreiser's characters display simultaneously the anxiety about their own social positions and the complacency about the class hierarchy itself that constitute Veblen's most familiar theme. Many of Dreiser's creations appear to be victims of the American class structure: Carrie Meeber of *Sister Carrie* (1900), Clyde Griffiths and Roberta Alden of *An American Tragedy* (1925), and the title character of *Jennie Gerhardt* (1911) all mistake class anxieties for personal inadequacies; Frank Cowperwood of *The Trilogy of Desire* (*The Financier* [1912], *The Titan* [1914], *The Stoic* [1947]) is perceived by other characters as a pariah despite (and because of) his immense wealth; Eugene Witla, the hero of Dreiser's autobiographical *The "Genius"* (1915), can barely paint because he agonizes so over money and status. Yet rebelling against the class system could not be further from these characters' minds. While individual characters experience frequent and, at times, dramatic changes in their social status, the class hierarchies remain intact. Dreiser and Veblen discerned how the United States class system replicates itself through the cooperation of its members, in spite of the mythically resonant cases of social mobility, and in spite of the damage the class system causes many individuals. Dreiser and Veblen consistently expose American attitudes about class as examples of what might be called an instinct of denial.

The interdisciplinary nature of Veblen's and Dreiser's cultural criticism also illuminates some striking exchanges among social science and realistic fiction during this period. It is not coincidental that Dreiser's novels made a lasting impact on several "Chicago school" sociologists, such as Robert Park and Ernest Burgess—nor that Veblen's stepdaughter recalls his declaring, in the last year of his life, that he wished he could have written "just one good novel." I am interested in both the social scientific aspects of Dreiser and the rhetorical dimensions of Veblen. Dreiser demonstrates what C. Wright Mills describes as "the sociological imagination," and this imagination led him to cultural criticism.[8] Similarly, Veblen's rhetorical maneuvering, especially as seen in his satire, continually edges his social theory toward cultural criticism. Reading Veblen through Dreiser, and Dreiser through Veblen, I want to show how literature and social science can merge in cultural criticism.

Dreiser, Social Science, and Cultural Criticism

As the nearly simultaneous publications of their watershed works, *Sister Carrie* (1900) and *The Theory of the Leisure Class* (1899), may suggest, I will not be arguing for direct influence in either direction. There are, however, indications of Dreiser's familiarity with Veblen's works. For instance, the Pennsylvania edition of *Newspaper Days* restores a paradigmatic Dreiser moment when he describes an internal debate over the attractions of several young women. Of the wealthier one, Dreiser recalls that he "made invidious comparisons from a material point of view and wished that I could marry some really wealthy girl" (*ND*, 88). Not only is "invidious comparison" one of Veblen's most memorable phrases, it always proceeds from a "material point of view" in his social theory. In a 1935 letter to Donald McCord (the

8. Carla Cappetti, *Writing Chicago: Modernism, Ethnography, and the Novel*, 24–25, 28; Becky Veblen Meyers, August 20, 1964, letter to Barbara Kevles in Carleton College Archives (Thorstein Veblen Collection); C. Wright Mills, *The Sociological Imagination*. Mills defines the sociological imagination as "the capacity to shift from one perspective to another—from the political to the psychological; from examination of a single family to comparative assessment of the national budgets of the world; from the theological school to the military establishment; from considerations of an oil industry to studies of contemporary poetry. It is the capacity to range from the most impersonal and remote transformations to the most intimate features of the human self—and to see the relations between the two" (7).

In "Fiction and the Science of Society," Susan Mizruchi describes the sociologist as a "professionalized" version of a novelist, both engaged in "social observation, description of human types and types of interaction, [and] the classification of these types" (191). And in *Writing Chicago,* Cappetti shows how "Before 'disciplinary' divisions separated them, personal, artistic and political ties joined the sociologist and the novelist" (16). Other discussions of literature and the natural or social sciences (a field that has grown exponentially in recent years) relevant to my argument include Gillian Beer, *Darwin's Plots: Evolutionary Narrative in Darwin, George Eliot, and Nineteenth-Century Fiction;* Wolf Lepienies, *Between Literature and Science: The Rise of Sociology;* George Levine, "By Knowledge Possessed: Darwin, Nature, and Victorian Narrative," *One Culture: Essays in Science and Literature,* and (as editor) *Realism and Representation: Essays on the Problem of Realism in Relation to Science, Literature, and Culture.*

original for "Peter" in *Twelve Men*), Dreiser describes *Leisure Class* as "marvellous."[9] Dreiser even acknowledges that a Veblenian viewpoint permeates *An American Tragedy* (1925) when he describes the novel as illustrating the development of "our 'leisure class,' " the proprietary quotation marks indicating one author's respect for a phrase that had become indelibly associated with another (*SUP,* 292). Dreiser seems also to have been aware of Veblen's notoriety, much like his own, as a womanizer. In a 1920 diary entry, he records meeting Florence Deshon (the model for *A Gallery of Women*'s "Ernestine") who reported "The visit of Thorstin Veblin [*sic*] and the second hit she made, on account of her beauty I presume" (*AD,* 349). In sum, Dreiser was familiar with some of Veblen's ideas and also with his reputation.

But the connection between these two figures runs deeper than can be accounted for by direct influence. As midwesterners whose parents had emigrated from Europe (Dreiser's father from Germany in 1844, Veblen's parents from Norway in 1847), they were products of similar backgrounds.[10] While biographical similarities no doubt account for some of the congruity in their viewpoints, I would emphasize instead that the ideas of Veblen—Dreiser's senior by fourteen years—pervaded intellectual discussion in the first quarter of this century. In a 1918 letter to the *Dial,* Maxwell Anderson suggests a climate of Veblen-by-osmosis: "I once asked a friend if he had read *The Theory of the Leisure Class.* 'Why no,' he retorted, 'why should I? All my friends have read

---

9. Dreiser to Donald P. McCord, September 10, 1935, in *Letters of Theodore Dreiser,* 2:750.

10. The standard biography, Dorfman's *Thorstein Veblen and His America,* mistakenly describes the family as poor and socially marginal. During Veblen's childhood, his parents owned two hundred ninety acres of prime Minnesota farming land. Thorstein Veblen's older brother, Andrew, corresponded extensively with Dorfman, warning him of his errors well before the biography was published. In his 1968 introduction to *Thorstein Veblen: The Carleton College Veblen Seminar Essays,* Carlton C. Qualey drew attention to Andrew Veblen's warnings (3–4). John Kenneth Galbraith issued an extensive rebuttal to the Dorfman-inspired portrayal of Veblen in his 1973 "A New Theory of Thorstein Veblen."

Recent reconstructions of Veblen's biography include Rick Tilman, *Thorstein Veblen and His Critics, 1891-1963: Conservative, Liberal, and Radical Perspectives,* chap. 1, and *The Intellectual Legacy of Thorstein Veblen: Unresolved Issues,* chap. 1; Stephen Edgell, "Rescuing Veblen from Valhalla: Deconstruction and Reconstruction of a Sociological Legend"; Ralf Schimmer, *Populismus und Sozialwissenschaften im Amerika der Jahrhundertwende* and "Wider die Legende von der unüberbrückbaren Distanz: Der amerikanische Populismus als normativer Gehaltgrund der Veblenschen Sozialkritik"; Russell H. Bartley and Sylvia E. Bartley, "In Search of Thorstein Veblen: Further Inquiries into His Life and Work."

it. It permeates the atmosphere in which I live.' "[11] Veblen's influence during Dreiser's lifetime was so pervasive that Dreiser could similarly have "read" him without ever opening a single book.

In the poem "Vital Statistics," an evocative portrait of Dreiser, Robert Penn Warren imagines the novelist reading, and in his behaviors illustrating, Veblen's books. Exercising considerable poetic license in suggesting that Dreiser pored over Veblen as he did over Freud, Warren accurately captures Dreiser's intellectual and personal longings.[12] I would modify Warren's sketch only in suggesting that Veblen's ideas do, in a sense, "help" Dreiser—or more accurately, they help us understand his sociologically informed, economically astute analysis of American culture.

Dreiser's self-described "abnormally inquiring mind" (*D*, 32) led him to study the social as well as the natural sciences and to adopt their insights in his writing.[13] Lacking Veblen's multiple academic degrees, Dreiser's interest in science was largely compensatory for his meager formal education. In the autobiography of his earliest years, *Dawn* (1931), Dreiser says that he and his siblings lacked any "social or constructive sense" and consequently "mooned and dreamed and fiddled away their time" (108). The phrasing reveals Dreiser's belief that effective agents must comprehend social processes. He consequently berates himself for having "no gift for organic sociology" (*D*, 326-27), yet he attempts to become a sociologist in such works as *Dreiser Looks at Russia* (1928), *Tragic America* (1931), *A Hoosier Holiday* (1916), and many of the essays in *Hey Rub-a-Dub-Dub* (1920). Dreiser's entire body of work—perhaps especially the fiction—reveals his groping for a way to make sense of the capitalist socioeconomic order. In Dreiser's often-repeated words, he desperately wanted to figure out "how life was organized" (*F*, 7; *D*, 10; *JG*, 194). His fascination with how Herbert Spencer believed life to be organized is an established fact, but Veblen's social theory—more insightful and enduring than the famous Social Darwinist's—finally explains Dreiser's mind and work better, especially his attempts to analyze society critically.

11. Maxwell Anderson quoted in Caraher, "Thorstein Veblen," 2.
12. Robert Penn Warren, "Vital Statistics," in *Homage to Theodore Dreiser*, 6.
13. On Dreiser's interest in science, see Ellen Moers, *Two Dreisers*, especially 133-69; Ronald Martin, *American Literature and the Universe of Force*, chap. 7; Robert H. Elias, *Theodore Dreiser: Apostle of Nature;* and Louis J. Zanine, *Mechanism and Mysticism: The Influence of Science on the Thought and Work of Theodore Dreiser.*

Dreiser's musing over social organization led him to conclude that conventions and institutions, which only masqueraded as transcendent and necessary rules, regulated human conduct. Dreiser rebelled at this discovery. His sociological imagination led him to crusade against what he and his ally H. L. Mencken termed "puritanism" in literary, social, and sexual mores. Both the substance and the style of these crusades established Dreiser as a formidable cultural critic for his generation, but this dimension of his work has lately been obscured. Recently, critics have stressed instead the complicity of Dreiser's works with capitalism.[14] This emphasis provides a welcome corrective to some earlier treatments that overstated Dreiser's "distance" from or "opposition" to his culture, but it disregards his self-perception and reputation during his lifetime as an imposing cultural critic.

The cultural criticism can admittedly be difficult to disentangle from the novelist's glorification of some of the showier manifestations of capitalism (such as nubile starlets, beautiful heiresses, and virile financiers). Such contradictions permeate Dreiser's viewpoint as a novelist and as a critic, surfacing even in some of his self-descriptions, where one would presume he would be especially careful. In *Dreiser Looks at Russia* (1928), for instance—an account as intensely self-conscious as it is society-conscious, and a work that engages in sustained criticism of both the Soviet Union and the United States—Dreiser at one point describes himself as an observer reporting on the Soviet experiment "to this outer world of benighted capitalism of which I am still, I lament, a somewhat troubled part" (*DLR,* 203).

If Dreiser himself knows he is a part of what he criticizes, then asking whether he is "for" or "against" capitalism is simply, as Walter Benn Michaels has argued, the wrong question. As Michaels puts it in a frequently quoted passage, "Dreiser didn't so much approve or disapprove of capitalism; he desired pretty women in little tan jackets with mother-of-pearl buttons, and he feared becoming a bum on the streets of New York. These fears and desires were themselves

---

14. Kaplan, *Social Construction,* 7. Literary realists have received much bad press recently for allegedly supporting the status quo. For instance, according to Leo Bersani's influential *A Future for Astyanax: Character and Desire in Literature,* "The formal and psychological reticence of most realistic fiction makes for a secret complicity between the novelist and his society's illusions about its own order" (62–63). In *American Literary Realism and the Failed Promise of Contract,* Brook Thomas provides a welcome critique of what he terms the "police academy approach to realism" (14).

made available by consumer capitalism." Michaels's claim is undeniable: Dreiser, like all other observers, cannot presume to occupy a space outside of his culture from which he can analyze it. But Michaels's point has been misinterpreted; he is not discrediting either the analysis or the criticism of cultures. Helping to place cultural analysis on a new foundation, Edward Said has recently defined the intellectual as a person necessarily both public and private. Using Sartre as his example, Said argues that the "complications" of one's personal circumstances, "[f]ar from disabling or disqualifying him as an intellectual, . . . give texture and tension to what he said, expose him as a fallible human being, not a dreary and moralistic preacher."[15] Recognizing that all writers and critics are implicated in the culture they examine allows us to see that Dreiser's personal investment in his subjects only enhances the authority, durability, and interest of his cultural criticism.

As an astute analyst of the American scene, Dreiser often beats us to the punch line. As Philip Fisher—to my knowledge the first critic to notice this quality of Dreiser's—puts it, his novels transform what was previously "unimaginable" into "the obvious."[16] To mention only two of Dreiser's prescient concerns, he is, notwithstanding his notorious sexual "varietism," extremely attuned to the problem of pregnancy for young, unmarried women and the differential access to both birth control and abortion based on their class positions. And seventy years before Americans gorged themselves on the O. J. Simpson trial, Dreiser sharply (and deservedly) criticized our fascination with crime and punishment, as seen, for instance, in his trenchant critique of the selling of Clyde Griffiths and Roberta Alden—along with peanuts and other snacks—during the lengthy trial sequence of *An American Tragedy.*

Yet more than the subjects Dreiser addresses, it is his intellectual orientation that reveals affinities with the practice of cultural criticism today. He anticipated the recent autobiographical turn in cultural studies, liberally borrowing from personal and familial history to construct novels as well as nonfiction and aggressively using personal experiences in his writings. As Elizabeth Kearney, who was one of Dreiser's assistants and may have been one of his lovers, remarks in an unpublished

---

15. Michaels, *Gold Standard,* 19; Edward Said, *Representations of the Intellectual,* 14. I am grateful to Larry Van Sickle for alerting me to the relevance of Said's book to Veblen in an unpublished paper, "The Pathologizing of Thorstein Veblen: He Ain't No Lord Keynes But He Just Might Be Redemptive."

16. Fisher, *Hard Facts,* 8.

interview, "It is impossible to dissociate Theodore Dreiser from his works."[17] Dreiser's relentless blurring of fact/fiction, private/public, self/other, which occurs throughout his writings, anticipates the position assumed by many cultural critics today.

Dreiser also has much to say about the now-popular topic of the role of the public intellectual in the United States. For instance, in an unpublished essay, "The Professional Intellectual and His Present Place," Dreiser goes after thinkers who fail to see their responsibility to the public. He indicts those who won't engage contemporary problems, arguing that would-be intellectuals cannot "pose as scholars and leaders in matters of mind" without committing themselves politically; they must be critical intellectuals to be intellectuals at all. Dreiser then asks a question that still resonates: "What power have they [the professional intellectuals]?"[18] His own quest for power through the development of a critical mind led him to an epiphany. "Books! Books! Books!" he writes in *Dawn*, "I was reading and awakening to a consciousness of many things, the mere knowledge of which appeared to coincide with power. . . . One could do things with sufficient power" (200). As arresting as Dreiser's equation of power with knowledge—a conjunction increasingly popular with critical intellectuals since Foucault—is his confidence that effective action can result from joining the two.

Veblen, Stylistics, and Cultural Criticism

Although this study began as an interpretation of Dreiser's works as read through Veblen, I increasingly found that Veblen's social and economic theory not only demanded interpretation, but also became clearer the more I implicated it with Dreiser. Veblen's attempts "to eliminate the 'personal equation' " (*AO*, 262) from his writing contrast sharply with Dreiser's autobiographical tendency, but his cultural criticism also anticipated many modernist, even postmodern, attitudes.[19] Veblen has much to contribute to current concerns over the tendency of theory

---

17. Folder 13400, Theodore Dreiser Collection. Van Pelt Library, University of Pennsylvania.
18. Folder 12263, Dreiser Collection.
19. On early twentieth-century social sciences and modernism, see Dorothy Ross's collection, *Modernist Impulses in the Human Sciences, 1870–1930;* also see Hartwig Isernghagen, " 'A Constitutional Inability to Say Yes:' Thorstein Veblen, the Reconstruction Program of *The Dial,* and the Development of American Modernism after World War I."

to make intellectuals simultaneously vain and petty, while obscuring facts that fail to correspond with theoretical expectation. Veblen's thought was pragmatic in that he denied any theoretical construct could provide access to transcendent, fixed truth; at the same time, he believed that dismantling existing theories could help people to see more clearly.[20]

By presenting Veblen as a multifaceted cultural critic whose style and subject are inseparable, I want to help resolve several problems that beleaguer studies of his work. All too often, commentators examine an isolated theme (such as conspicuous consumption or the cultural role of the engineer) or disciplinary perspective, which results in a truncated view of Veblen. This problem is exacerbated by the fact that many readers do not venture beyond *Leisure Class,* which certainly does not contain everything Veblen had to say, as H. L. Mencken once claimed in a hilarious rebuke.[21] In typically indirect fashion, under the cover of discussing another thinker, Veblen explains how to read his own books: the only way to grasp any of Marx's ideas is to understand them all, for to look at an "isolated feature" of Marxism from the standpoint of an orthodox theoretical perspective "is as futile as a discussion of solids in terms of two dimensions" (*POS,* 410). Similarly, the only way to grasp the full implications of Veblen's ideas is to trace them through all of his writings. His beliefs also become clearer when examples are generated (Veblen did not provide them as often as one might wish), and so I use Dreiser to furnish concrete illustrations.

Among those who note Veblen's relevance to postmodernism are Warren J. Samuels, "The Self-Referentiability of Thorstein Veblen's Theory of the Preconceptions of Economic Science" and introduction to *The Place of Science in Modern Civilization and Other Essays* by Thorstein Veblen; Teresa Toulouse, "Veblen and His Reader: Rhetoric and Intention in *The Theory of the Leisure Class*"; Doug Brown, "An Institutionalist Look at Postmodernism"; and John P. Diggins, *The Promise of Pragmatism: Modernism and the Crisis of Knowledge and Authority,* 468–69.

20. While at Johns Hopkins, Veblen studied logic under Charles Saunders Peirce. It is always difficult to trace Veblen's relation to existing intellectual traditions, but if we follow Rorty's definition of pragmatism in *Consequences* as "simply anti-essentialism applied to notions like 'truth,' 'knowledge,' 'language,' 'morality' " (162), then Veblen is clearly a fellow traveler. Yet Veblen also differs from pragmatism, for instance, in maintaining a considerably less instrumentalist idea of what one should do *with* knowledge. See also William T. Waller Jr., "The Concept of Habit in Economic Analysis"; Diggins, *The Promise of Pragmatism;* E. E. Liebhafsky, "The Influence of Charles Saunders Peirce on Institutional Economics"; and Alan W. Dyer, "Veblen on Scientific Creativity: The Influence of Charles S. Peirce."

21. H. L. Mencken, "Professor Veblen." Mencken later remarked in a letter to Joseph Dorfman that he never intended his lambaste to be taken seriously (cited in Dorfman, "New Light," 20–21).

Veblen scholarship has also been inhibited by the unease most commentators feel about one of his characteristic rhetorical modes: satire. John Dos Passos astutely remarked, "People complained they never knew whether Veblen was joking or serious." (Dreiser's detractors have implied that he never knew when he was being funny, either.) The ambiguity of Veblen's acerbic humor has restrained discussion of his satire, and scholars have overlooked how purposeful it is.[22] Far more than a stylistic idiosyncrasy, Veblen's satire enacts a critique of one of his favorite targets: ceremonialism.

Veblen's concept of ceremonialism anticipates some of the outlines of Jean Baudrillard's characterization of the postmodern "age of simulation," which "substitute[s] signs of the real for the real itself." The images generated by consumer culture, according to Baudrillard, assume a life of their own, often towering over the objects to which they refer. One thinks immediately of those huge golden arches—how they dwarf the paltry burgers they advertise. Although Veblen has much to say about the importance of image, his idea of ceremonialism emphasizes the displacement of reality through ritual. According to Veblen, ritualistic constructs such as status and "make believe" events such as sports and military parades function as ceremonial surrogates for reality. In contrast with behaviors that Veblen favors for their utility—such as building a shed or planting a garden—ceremonialism endangers human welfare by confusing people about what is important. We worship at the false altar of the ceremonial, praising the useless and looking down on the useful. (Ceremonial behaviors emerge throughout Dreiser's novels, from the mildly self-destructive ritualistic drinking at Fitzgerald and Moy's saloon in *Sister Carrie* to the far more insidious quest for pseudojustice in book three of *An American Tragedy*.) The importance of ceremonialism to Veblen's thinking has been duly noted, but commentators have not realized how his satire exposes a ceremonial distinction between the serious/profound and the comic/quotidian. Far more than a game or a joke, Veblen's satire does serious work. William James remarked of *Leisure Class* that it was "awfully jolly in

---

22. John Dos Passos, "The Bitter Drink," 97. Important discussions of Veblen's satire include Daniel Aaron, *Men of Good Hope: A Story of American Progressives;* Toulouse, "Veblen and His Reader"; and Stephen Conroy, "Thorstein Veblen's Prose."

A small group of economists insists on the rhetorical nature of their field, including Donald N. McCloskey, *The Rhetoric of Economics;* Hodgson, *Economics and Evolution,* especially chap. 2; and William Waller and Linda R. Robertson, "Why Johnny (Ph.D., Economics) Can't Read: A Rhetorical Analysis of Thorstein Veblen and a Response to Donald McCloskey's *Rhetoric of Economics.*"

spots, & telling much truth" and, in fact, Veblen discloses the most startling revelations while making us laugh. Veblen demands that we take laughter seriously (but never solemnly), that we apprehend what he called "The Science of Laughter."[23]

Veblen's distinctive fusing of social theory and cultural criticism through satire exemplifies what M. M. Bakhtin describes as "double voicedness." One of the most striking reactions to Veblen's double voicedness can be seen in the reception of his *Imperial Germany and the Industrial Revolution* (1915), which was simultaneously praised by the Committee on Public Information for its anti-German sentiments—and banned from the mails by the postmaster of New York City for being sympathetic to the German cause. While Veblen could scarcely have predicted those conflicting interpretations of his book, he must have relished how they made the government agencies appear absurd. More typically, Veblen's double voicedness generates satire. Speaking out of two (or more) sides of his mouth, he can declaim in an ostensibly serious tone a "scientific" point, while deriding whatever satiric object presents itself. Veblen's satire teaches his readers how to talk back to institutions and laugh their way through ceremonialism. As Bakhtin explains the liberating process, laughter can elude "official" views of reality and thereby "remai[n] outside official falsifications, ... uninfected by lies."[24] Through satire, Veblen uses laughter to make ceremonial behaviors appear idiotic and thus to clear room for new ideas.

The satiric bite of Veblen's language makes him eminently quotable; many of the phrases he popularized, such as "conspicuous consumption" and "trained incapacity," are more widely recognized than his name. But Veblen's lifelong interest in rhetoric, language, and literature is not just an amusing quirk; it is central to the meaning of his works. Considerable external evidence confirms Veblen's acute sensitivity to language. Student lore at the University of Missouri had him

---

23. Jean Baudrillard, "The Precession of Simulacra," 254; William James, *Selected Unpublished Correspondence, 1885-1910*, 241-42. "The Science of Laughter" was the title of one of Veblen's college rhetoricals (see Dorfman, *Thorstein Veblen*, 32). Despite a tendency to impute to Veblen far more reformist tendencies than he in fact manifested, Chris Rojek's "Baudrillard and Leisure" provides a balanced discussion of the similarities and differences between Veblen and Baudrillard. For a useful discussion of ceremonialism, see Paul D. Bush, "The Theory of Institutional Change."

24. M. M. Bakhtin, *The Dialogic Imagination*, 371, 236. Dorfman discusses the reception of *Imperial Germany and the Industrial Revolution* in *Thorstein Veblen*, 382.

knowing twenty-six languages (no doubt an exaggeration), but Veblen read extensively and indisputably followed literary trends.[25] Edward Bellamy's utopian novel *Looking Backward* (1888) made a great impact on Veblen and his first wife, Ellen Rolfe, herself a creative writer. In the preface of the first book Veblen completed, a translation of an Icelandic saga into English, he muses over the challenge of translating idiomatic dialogue (Veblen's *Laxdæla Saga* was not published until 1925).[26] Although no direct evidence has surfaced that Veblen read Dreiser, he knew the works of authors associated with him, for he alludes to Edgar Lee Masters and Sinclair Lewis in *Absentee Ownership* (1923), describing the country town as "the perfect flower of self-help and cupidity standardised on the American plan. Its name may be Spoon River or Gopher Prairie" (142). Veblen's interest in literature and language informs his viewpoint, much as Dreiser's fascination with social science pervades his orientation.

One of the greatest obstacles to comprehending Veblen has been his defiance of such epistemological boundaries. A fascinating response to his capacious intellect comes from John Cummings, an economist who reviewed *Leisure Class* and to whom Veblen uncharacteristically responded in print.[27] Cummings eventually came to see that *Leisure*

25. Jacob Warshaw, Veblen's colleague at the University of Missouri, records the student legend in an unpublished essay, "Recollections of Thorstein Veblen." This document seems to be the same as an unpublished paper that Rick Tilman cites, in *Thorstein Veblen and His Critics,* as "A Few Footnotes to Dorfman's Veblen" (both held by the Western Historical Manuscript Collection, University of Missouri–Columbia). I refer to a photocopy of "Recollections" held in the Joseph Dorfman papers at Columbia University.

Like the rest of his family, Veblen closely followed the work of Bjornstjerne Bjornson, the Norwegian playwright, novelist, poet, and 1903 Nobel Laureate. Veblen also delighted in Ibsen, and according to a student at the University of Chicago, he often quoted Chaucer and Cervantes (Dorfman, *Thorstein Veblen,* 43, 119). The years Veblen served as an editor for *The Dial* further suggest his knowledge of literary trends.

The recently discovered library of Veblen from his Wisconsin cabin contains not only Darwin, Loeb, Spencer, Ricardo, and Marx but also Shakespeare, Swift, Dante, Carlyle, Balzac, Shaw, Hardy, Knut Hamsun, and a novel by Jack London (*The Iron Heel*). On this library, now housed at Carleton College, see Russell H. Bartley and Sylvia Yoneda, "Thorstein Veblen on Washington Island: Traces of Life."

26. The Wisconsin Historical Society holds seven unpublished letters from Veblen to the distinguished Scandinavian scholar Rasmus Bjorn Anderson that confirm he had completed a translation of the saga by 1890 while living with his wife Ellen at Stacyville, Iowa, on property owned by her father. These letters tell the story of Veblen's first (unsuccessful) attempt to publish a book. Not until nine years later would he publish *Leisure Class.*

27. See Veblen, "Mr. Cummings's Strictures on 'The Theory of the Leisure Class.'"

Class had baffled him because it "was new art—modernism of the most bizarre sort in economic writing and thinking, and in my hidebound conventionalism, my reaction to it was much the same as my reaction to some of the more extreme forms of ultra-modernism in poetry, music, or painting. Veblen was cubistic and for the time being incomprehensible." Thus it took Cummings many years, by his own confession, to comprehend what he called Veblen's "jazz economics."[28] Veblen, no doubt, would have considered Cummings's distinction between social science and music yet another ceremonial one.

According to Rick Tilman, "No consensus now exists on the value or even the meaning of Veblen's work."[29] Perhaps the only area of agreement is that Veblen's work defies categorization. Much of the conundrum about "placing" Veblen derives from the absence of a consistent methodology (as method is usually conceived) in his work and his related contempt for systematic thinking. Veblen considered the building of all-encompassing theories, now called "metanarratives" or "regimes of truth," to be the bane of social science. Veblen's interdisciplinary form of cultural criticism avoids metanarrative in favor of a series of case studies. The most memorable include business as sabotage, women as vicarious consumers for their fathers and husbands, patriotism as dementia, and academia as for sale to the highest bidder. When taken together, Veblen's case studies illustrate an immensely useful vantage point from which one can examine countless social phenomena and cultural productions. Veblen's distinctive method of critical analysis needs to be understood as a *style*. Like other distinctive and successful styles—whether Hemingway's prose, Sullivan's skyscrapers, or Cézanne's landscapes—Veblen's style is unmistakable.[30] It is also eminently suited for its intended function: cultural criticism.

This study proceeds in the spirit of the grand conversation among the disciplines advocated by Richard Rorty. But as the host of any dinner party knows, conversation is most stimulating when participants voice differences of opinion as well as agreement. While the analogies between Dreiser and Veblen form my primary subject, I do

28. John Cummings, February 5, 1935, letter to Joseph Dorfman, Joseph Dorfman Collection, Thorstein Veblen correspondence, "C" file, Butler Library, Columbia University.
29. Tilman, *Thorstein Veblen and His Critics*, 13.
30. See also, Christopher Shannon, *Conspicuous Criticism: Tradition, the Individual, and Culture in American Social Thought, from Veblen to Mills*, 1.

not intend to forget that—as one of America's finest sociological novelists, James Baldwin, remarked—"literature and sociology are not one and the same."[31] By addressing some areas of theoretical confrontation between Dreiser and Veblen in the chapters that follow, I illustrate differences between their critical perspectives along with the ground that they shared. The first and fourth chapters emphasize, respectively, the analogous rhetorical and axiological premises informing their cultural criticism. The second and third chapters take up Veblen's and Dreiser's examinations of identical subjects, about which they draw different conclusions. While the sequence of chapters conveys an unfolding argument, they may be read in any order; I have endeavored to write each chapter so that it can stand alone.

Chapter 1 focuses on how a similar notion of cultural criticism determines the rhetorical strategies shared by Veblen and Dreiser through a wide variety of writings. They position themselves as authoritative cultural critics by defining themselves as lonely truth tellers speaking through a Babel of lies; attacking privileged sources of authority such as the church; and constructing a countermyth of personal intransigence. I trace these components of their rhetoric of confrontation through Veblen's attacks on orthodox economic theory in numerous essays and books, also taking up his discussion of "idle curiosity" and the cultural role of the university; and through Dreiser's autobiographies and other nonfiction, along with his most autobiographical novel, *The "Genius."*

Chapters 2 through 4 then target discrete subjects that recur throughout Veblen and Dreiser's cultural criticism and proceed by placing some of Veblen's key social scientific concepts into dialogue with Dreiser's novels. Each of these chapters emphasizes a different area of social science that Veblen brings to bear on cultural criticism: economics in chapter 2, psychology in chapter 3, and anthropology in chapter 4.

Veblen's distinction between business and industry, developed especially in *The Theory of Business Enterprise* (1904) and critical in both senses of the word, forms the center of chapter 2. I use Veblen's analysis to account for the oddly insubstantial nature of business that Dreiser depicts in *The Trilogy of Desire* and to examine persistent complaints about his supposedly inconsistent and naive perspective on

---

31. Richard Rorty, *Philosophy and the Mirror of Nature*, 389–94 (see also Rorty, *Consequences,* xliii); James Baldwin, "Everybody's Protest Novel," 31.

capitalism. Yet while Veblen and Dreiser similarly analyze the operation of speculative finance, they conflict in their evaluations of the sort of cultural work performed by business. Veblen considers business "immaterial" in having literally nothing behind it; he further criticizes it as a socially and politically reactionary force. Dreiser, however, reads the immateriality of finance to mean transcendence of conventional pieties and institutional restraints. And, much to the disbelief or incomprehension of most critics, Dreiser represents the *Trilogy*'s hero, Frank Cowperwood, as an agent of revolutionary change—in fact, as something of a cultural critic himself.

Chapter 3 takes up the impact of capitalism on human psychology, as observed through the behavior of consumers. Focusing on Veblen's psychological model, particularly the concepts of "pecuniary emulation" and "invidious comparison" associated with *Leisure Class* and explored throughout his writings, this chapter analyzes the distinctive methods of self-construction seen in Dreiser's two most famous novels. *Sister Carrie* and *An American Tragedy* demonstrate how identity formation under capitalism results in the perpetuation of a social order that oppresses and even enslaves citizens. In his first novel, Dreiser tends to accept without question the psychology critiqued by Veblen; in *An American Tragedy*, however, he moves to a more Veblenian perspective and denounces the psychology of desire.

Chapter 4 turns to the "industry" half of Veblen's industry/business dichotomy, focusing on the anthropological assumptions that underwrite his cultural criticism. These assumptions include essentialist views of women's work—a controversial issue then and now—and form the area of most complete axiological congruity between Veblen and Dreiser. Veblen's quasi-anthropological constructs, the "instinct of workmanship" and the closely related "parental bent," embody the unceremonial qualities he most values, but neither the feminine character of these "instincts," nor their function in Veblen's cultural criticism, has been sufficiently explored. The ethical position that allows Veblen to link workmanship with the parental bent corresponds precisely with that in Dreiser's *Jennie Gerhardt*. Like many feminists and protofeminists from the mid-nineteenth century through the early years of the twentieth, Dreiser and Veblen align women's work with motherhood in order to criticize the values of capitalist America. Thus *Jennie Gerhardt*, traditionally written off as one of Dreiser's softest and most sentimental novels, in fact constitutes one of his sharpest examples of cultural criticism, while the instinct of workmanship provides the ethical center on which Veblen grounds his entire critical practice.

One of the best descriptions of Veblen bills him as an "epic novelist"; one of the most astute characterizations of Dreiser notes that he provides as much a "criticism" as an "interpretation of life."[32] Veblen and Dreiser are indeed large; they contain multitudes. By bringing Dreiser to bear on Veblen (and vice versa) I want to liberate their works from critical straitjackets. The rubric of cultural criticism provides a way to open up Dreiser and Veblen to the inter- and cross-disciplinary analysis they both practiced, as well as to present debates about the role of the critical intellectual.

---

32. Max Lerner, introduction to *The Portable Veblen* by Thorstein Veblen, 46; H. L. Mencken, "A Novel of the First Rank," 741.

1

The Rhetoric of Confrontation

> "The sum and substance of literary as well as social morality may be expressed in three words—tell the truth. It matters not how the tongues of the critics may wag."
>
> —*Theodore Dreiser*

> "Skepticism is the beginning of science."
>
> —*Thorstein Veblen*

The durability of any institution, as well as its power to influence the thoughts and actions of individuals, depends upon the authority that underwrites it. Thus, as anthropologist Mary Douglas explains, "most established institutions, if challenged, are able to rest their claims to legitimacy on their fit with the nature of the universe." This institutional ruse of permanency and inevitability infuriates Dreiser and Veblen. They consider the institutions most people take for granted to be historically conditioned and, thus, neither inevitable nor permanent, but tentative and fallible. Their focus on discrediting authorities believed to be "natural," "normal," and "eternal" by insisting on the historical formation and evolution of institutions anticipates Edward Said's recent definition of the twentieth-century intellectual as one who "see[s] things not simply as they are, but as they have come to be that way."[1] In the process of showing how powerful institutions such as business

---

1. Douglas, *How Institutions Think*, 46; Said, *Representations*, 60.

enterprise and Christian morality have come to be, Veblen and Dreiser also attempt to shift the locus of authority from the institutions in which power has become entrenched to themselves as cultural critics.

The cultural criticism of Veblen and Dreiser is both a cause and an effect of assuming the paradoxical position of the "radical with authority," to borrow John Diggins's apt phrase.[2] The problem of establishing one's authority after challenging the whole notion of authority has only become more familiar in recent years, as an increasing number of intellectuals (and their critics) conceive of intellectual work as antagonistic toward received values. Challenging institutionalized power while building up a contrary position of authority, Dreiser and Veblen employ a similar rhetoric of confrontation, and this chapter traces three of its key elements. Two of these tactics depend on aligning their critical pronouncements with the authority of scientific discourse. First, by highlighting facts that they claim have been left out of preexisting accounts of reality, Dreiser and Veblen make other authorities seem biased while casting themselves as more credible analysts of the contemporary scene. Second, armed with a new sense of scientific truth, Dreiser and Veblen use empirical data to combat the authority of conventional morality (most pointedly, as invested in Christianity). But the third element in their rhetoric of confrontation belongs more to the realm of myth than science, and its legacy remains palpable today. Veblen and Dreiser forge a new authority for the cultural critic by creating mythically resonant personae as independent intellectuals who sabotage the status quo. It is particularly by romanticizing the role of the oppositional intellectual that Dreiser and Veblen legitimize their style of cultural criticism.

Works that focus on the role of the oppositional intellectual best illustrate these strategies. Dreiser's nonfiction is of particular interest, especially his autobiographies such as *Dawn* (1931) and *A Book about Myself* (1922)[3] and sociological proclamations such as in *Hey Rub-a-Dub-Dub* (1920). His semiautobiographical novel about a painter's struggle with convention, *The "Genius"* (1915), and statements regarding the "suppression" of his various works, also show Dreiser

2. John P. Diggins, "A Radical with Authority," 32.
3. To avoid confusion between the version of *Newspaper Days* published during Dreiser's lifetime and the "restored" text published by the University of Pennsylvania Press in 1991, I refer to the 1922 version as *A Book about Myself* and the Pennsylvania edition as *Newspaper Days*.

developing the persona of rebellious artist as cultural critic. Veblen's defiance is especially pronounced in his critiques of economic orthodoxy, such as in essays reprinted in *The Place of Science in Modern Civilization* (1919) and *Essays in Our Changing Order* (1934), and in his assault on academia in *The Higher Learning in America* (1918). Veblen attacks existing intellectual authorities to clear room for his own work in his celebrations of outsiders, including "The Intellectual Pre-Eminence of the Jews" (1919, in *ECO*) and *The Engineers and the Price System* (1921). While some of Dreiser's and Veblen's rhetorical strategies are unsuitable for today's cultural critics, the problems they faced remain substantially the same. Their refusal of moral or cultural givens, suspicion of foundationalist truth claims, and glorification of intellectual work as oppositional inaugurated a chapter in the history of United States cultural criticism that has not yet ended.

Telling the Truth

Edward Said's characterization of the intellectual's job as "speaking truth to power" accurately describes how Dreiser and Veblen envisioned their writing. With their first books, Veblen and Dreiser styled themselves as free thinkers working against the grain of America's dominant moral, political, and professional paradigms. The authority of their cultural criticism depends on the transformed sense of truth that they shared with sympathetic readers. For instance, Lester Frank Ward, often considered the father of American sociology, admired Veblen's first book, but he worried about how Americans would receive it. In his review Ward said that *The Theory of the Leisure Class* (1899) "contains too much truth" for most readers to swallow. One year later, the literary naturalist Frank Norris, who had recommended *Sister Carrie* (1900) for publication when he read it in manuscript, congratulated Dreiser on the appearance of "a *true* book in all senses of the word."[4]

We have learned to be wary of the epistemological baggage that goes along with truth claims, but the "truth" proffered by Veblen and Dreiser

---

4. Said, *Representations,* chap. 5; Lester Frank Ward, "The Theory of the Leisure Class," 619; Frank Norris to Dreiser, January 28, 1901, reprinted in the Norton Critical Edition of *Sister Carrie,* 454. Norris's views on literary *truth,* which he set as a higher goal than mere *accuracy,* are slippery; see, for instance, his essay "A Problem in Fiction," reprinted in *The Responsibilities of the Novelist.*

does not entail correspondence with some transcendent or unchanging reality. The role of the critical intellectual, as Veblen and Dreiser define it, is to tell the truth by accepting the Darwinian principle of ceaseless change, by resisting moralistic explanations of phenomena, and by observing things that have been invisible to others. In so defining truth, Dreiser and Veblen affirm the process of inquiry while insisting on the fallibility of knowledge.[5] They also present their critical orientation as a necessary corrective to deception and falsification.

Thus for both Dreiser and Veblen, telling the truth permits talking back to their own authoritative critics. In an early manifesto, "True Art Speaks Plainly" (1903), Dreiser defines his viewpoint as authentic and credible by pitting it against the lies perpetrated about his work. Writing in response to attacks on his "so-called immoral literature," Dreiser shifts the discussion from morality to epistemology and affirms "the business of the author . . . is to say what he knows to be true." This rhetorical contrast between truth (which he claims to obey) and morality (which he says blinds his critics) serves Dreiser in all his writing. He uses the same logic to distinguish his writing from the less confrontational literary realism of an earlier generation such as that of William Dean Howells. Dreiser implies that in critically challenging received wisdom, he will work a wider field: "The extent of all reality is the realm of the author's pen" he ambitiously claims, as if his predecessors had excluded parts of reality. Widening the canvas to take in previously ignored material helps to refute competing truth claims which, Dreiser insists, leave out too much of reality. He tautologically declaims, "Truth is what is; and the seeing of what is, the realization of truth."[6] But apprehending the whole truth is not, to judge by Dreiser's comments, easy for most of humankind. It falls to the cultural critic to see the invisible.

---

5. I draw these terms from David A. Hollinger's description of the pragmatist philosophers as affirming inquiry in the face of incomplete knowledge, in "The Problem of Pragmatism in American History," 30.

6. Theodore Dreiser, "True Art Speaks Plainly," 155-56. To be fair to Howells, Dreiser exaggerates his own originality. In *Criticism and Fiction,* Howells says the realist author "cannot look upon human life and declare this thing or that thing unworthy of notice, any more than the scientist can declare a fact of the material world beneath the dignity of his inquiry" (quoted in Michael Davitt Bell, *The Problem of American Realism: Studies in the Cultural History of a Literary Idea,* 32).

For an interesting treatment of the overlaps between literary and cultural criticism, see Morris Dickstein, *Double Agent: The Critic and Society.*

As to the wagging "tongues of the critics," Dreiser silences them by locating them within the ignorant past. With their assumptions of "immutable" aesthetic and moral values, Dreiser's "partially developed" critics reveal their limitations. Although they jealously guard "their own little theories," truth will out: "The impossibility of any such theory of literature having weight with the true artist must be apparent to every clear reasoning mind."[7] Dreiser turns the moralists' charge on its head: what they claim to be morally or theoretically wrong—such as Dreiser's own "true art"—he says is factually right.

Veblen makes an analogous declaration that true science speaks plainly, though he inverts Dreiser's concern with the self-evidence of facts in true literature to affirm the transparency of language in true science. Notwithstanding his own frequently loaded prose, Veblen declares, "No unprejudiced inquiry into the facts can content itself with anything short of plain speech" (*AO,* 156). And by offering his critical comments as instances of scientific "plain speech," Veblen protects himself from being criticized by others. For example, he accuses John Cummings, who had complained in his review of *Leisure Class* about the book's loaded language, of not understanding the meaning of plain words. According to Veblen, Cummings simply disliked his using "everyday words in their everyday meaning" (*ECO,* 30). Like Dreiser in "True Art Speaks Plainly," Veblen submits that his critics misconstrue his unguarded honesty: "It is an unhappy circumstance that all this [Veblen's] plain speaking . . . unavoidably has an air of finding fault" with theories that falsify reality (*AO,* 156).

Like Dreiser, Veblen claims to see more (and more accurately) than his competitors. In the preface to his first book, Veblen asserts that his unusual conclusions follow from his examining "everyday life" and "homely facts" rather than "recondite sources"; a similar disclaimer prepares the reader of his last monograph to find "workday facts" that will contradict "received theories" and "recondite information" (*TLC,* viii; *AO,* preface). Veblen's emphasis on the quotidian allows him to observe overlooked facts about formative institutions that shape people's perceptions of reality. Capitalism, for instance, is "so intimate a fact . . . that we have some difficulty in seeing it in perspective at all" (*POS,* 334). Veblen characterizes the problem as less the closeness of facts to the observer than the familiarity of theoretical formulations about them.

---

7. Dreiser, "True Art," 155–56.

By cultivating a distance from received interpretive schemes, Veblen purports to observe reality more accurately. For instance, he charges that popular descriptions of business enterprise, such as the notion of an invisible hand benignly regulating the market, neglect "obvious facts" to the contrary. Such accounts are "authenticated" only by "neatly avoid[ing]" any "inconvenient facts" and "'undesirable'" elements, such as anything "wasteful, disserviceable, or futile" (*POS*, 67; *ECO*, 156; *ECO*, 165 n, 164). Inconvenient facts such as the inequality and ruthlessness fostered by capitalism, Veblen satirically remarks, "have commonly been overlooked because they are too obvious to be seen except with the naked eye" (*ECO*, 115). In other words, to borrow a phrase from Mark Twain, most of us see capitalism through glass eyes, darkly. Veblen defines his own work as examining facts that have been invisible to others.[8]

Veblen and Dreiser's rhetoric of confrontation plays up a clash between two different notions of authority. In Dreiser's novel that deals most transparently with his sense of his own intransigence, *The "Genius,"* he depicts this clash of authorities as a war between two versions of truth: the conventional, respectable, but repressive certitude of society, and the iconoclastic, subversive, but liberating truth of the artist. Projecting onto Eugene Witla his idealized self-image as a confrontational intellectual, Dreiser registers the war between convention and truth in two related areas: Witla's artistic career and his struggle for sexual freedom.

Like the "homely facts" Veblen claims to champion, Witla's paintings depict "commonplace and customary" scenes "supposedly beyond the pale of artistic significance" (*G*, 235, 236). The key word is *supposedly*, for Witla's Ash Can painting distinctively renders subjects that other artists have ignored. His unusual perspective causes a favorable reviewer to declare Witla's exhibition true art that speaks plainly, with "no fear . . . , no bowing to traditions, no recognition of any of the accepted methods. . . . We have a new method" (*G*, 237–38). Dreiser forges this positive review out of his own aesthetic ideas, revealing how he tended to assume that addressing new subjects constituted formal innovation. But the reaction against Witla's work provides an equally important register of Dreiser's sense of artistic achievement. A conservative art publication condemns Witla for "insist[ing] on shabby

---

8. See T. W. Adorno's "Veblen's Attack on Culture" on Veblen's seeing the invisible.

details" and belittles his draftsmanship (*G,* 237). Dreiser again projects his experience onto the review of Witla's art; his own critics had already gotten considerable mileage out of mocking his plebeian subjects and inelegant style. The competing reviews represent sides in the debate over the nature of "true art" that Dreiser felt his own work had initiated and highlight the critics' tendency to underrate and misread his novels.[9]

Dreiser uses this battle to argue that critics of innovative art are moral and political reactionaries. In the splendid scene where a young Witla admires a Bouguereau nude at the Art Institute of Chicago, the narrator remarks that the painting was "anathema to the conservative and puritanical in mind, the religious in temperament, the cautious in training or taste" (*G,* 51). In fact,

> The very bringing of this picture to Chicago as a product for sale was enough to create a furore [*sic*] of objection. Such paintings should not be painted, was the cry of the press; or if painted, not exhibited. Bouguereau was conceived of by many as one of those dastards of art who were endeavoring to corrupt by their talent the morals of the world; there was a cry raised that the thing should be suppressed. (*G,* 52)

This alignment of received morality, conservative politics, and poor aesthetic judgment recurs throughout Dreiser's work. He asserts that ideology masquerades as aesthetics when censorious critics use their political and moral beliefs not only to discredit "true art," but also to deny the existence of anything that contradicts their notions of reality. Dreiser retaliates that these puritans, with their tyrannical morals and selective observation, attempt to censor not only art but truth itself. As he sees it, moralistic critics reveal their hypocrisy; they cry "Immoral! Immoral!" but "Under this cloak hide the vices of wealth as well as the vast unspoken blackness of poverty and ignorance."[10]

---

9. In a curious case of life imitating art, the controversy over publication of *The "Genius"* follows the terms Dreiser set out in the novel. Four hundred fifty-eight people signed a petition protesting the "suppression" of the novel due to lewdness and profanity. Many of the signatories agreed with H. L. Mencken about the dubious literary quality of the novel but defended it on the grounds of its author's truthfulness, affirming Dreiser's "manifest sincerity" ("A Protest against the Suppression of Theodore Dreiser's 'The Genius,'" 803).

On Dreiser's use of the careers of Ash-Can painters, see Joseph J. Kwiat, "Dreiser's *The 'Genius'* and Everett Shinn, the 'Ash-Can' Painter."

10. Dreiser, "True Art," 156.

Identifying Witla with a recognized artist, Bouguereau, Dreiser also tests out the idea of the persecuted, misunderstood artist central to his own persona as a cultural critic. Dreiser has to make the authority of the Bouguereau-Witla brand of sensuous realistic truth win out over the moralists' hypocritical and limited version. Dreiser's choice of Bouguereau, known for his sensuous, fleshy paintings, is strategic, for his argument in favor of Witla's notion of truth develops in the context of special pleading for promiscuity or, as Dreiser likes to call it, "varietism."

In one of the novel's most memorable scenes, Witla attempts to seduce his fiancée, Angela Blue, at her home in Blackwood. Angela offers to have intercourse with Witla, while pleading with him to turn her down—which, chivalrously, he does. Dreiser uses Angela's libidinal abandonment in this scene not to suggest her debauchery but rather the arbitrary nature of the codes regulating sexual conduct. When Angela worries that she hasn't been a "good girl," Witla responds (as would only a Dreiser hero) not to a coy mistress but as if to a philosophical protégée: "Right . . . is something which is supposed to be in accordance with a standard of truth. Now no one in all the world knows what truth is, no one." After this lecture, Witla dismisses Angela's conviction that good girls remain virgins until marriage as mere learned behavior and advances an argument regarding the cultural relativity of all beliefs, pompously citing Constantinople as an example (*G,* 129–30).

The position Dreiser advances in *The "Genius"* draws upon the romantic idea of the artist's transcendence of moral and social constraints, but it draws even more deeply on the Darwinian postulate of ceaseless change. Simply put, Dreiser champions Witla less because he paints gritty realistic scenes than because he utters relativistic Darwinian ideas and uses them to disclose the arbitrary nature of existing mores. As Eric Foner notes, evolutionary thought "became a major point of self-definition and self-justification for intellectuals" during this period. The application of Darwinian ideas to humans' social behavior is most familiar in the Social Darwinism of Herbert Spencer and his followers, a philosophy that deserves its bad reputation for racism, sexism, and generally flabby thinking. But as Carl Degler reminds us, there is nothing inherently conservative or regressive in the application of evolutionary principles to humans' social behavior. In fact, remarks Degler, turn-of-the-century reformers such as Lester Ward and Charlotte Perkins Gilman "could not help but find Darwinian evolution congenial since it clearly proclaimed the ubiquity and persistence of change."[11] Dreiser and Veblen were drawn to Darwin for precisely this reason. They use evolutionary principles not to rationalize the status quo, but to sabotage it, and their authority as cultural critics depends on adapting an evolving, provisional idea of truth.

Veblen describes the new sense of truth that characterized progressive intellectuals of his generation: "The question now . . . is not how things stabilise themselves in a 'static state,' but how they endlessly grow and change" (*ECO*, 8). That idea permeates *The "Genius,"* and Dreiser taxes the patience of readers by enlisting it to argue in favor of sexual promiscuity. Witla wonders, "Might one swear eternal fealty and abide by it when the very essence of nature was lack of fealty, inconsiderateness, destruction, change?" (*G*, 286).[12] A battle fought throughout the novel derives from Witla's desire to have sex with whomever he pleases before and after marrying Angela, a liberty that she resists granting him. But Witla's (and Dreiser's) self-serving question about monogamy points to the war that Dreiser is fighting for intellectual independence. He uses sexual license as a synecdoche for personal and intellectual freedom throughout his works, perhaps most pointedly in *The "Genius."*[13]

And to do Dreiser justice, he seems more interested in Angela's pre-Darwinian ideas than in her sexuality. She functions primarily as Witla's ideological foil, advocating static truths ("As it was written socially and ethically upon the tables of the law, so was it" [*G* 65)], while Witla affirms, in Darwinian spirit, that social orders evolve because of "accidental harmonies blossoming out of something which meant everything here to this order, nothing to the universe at large" (*G*, 118).

---

11. Eric Foner, introduction to *Social Darwinism in American Thought,* by Richard Hofstadter, xix; Carl Degler, *In Search of Human Nature: The Decline and Revival of Darwinism in American Social Thought,* 112.

12. In an essay on "Marriage and Divorce," Dreiser similarly remarks, "the trouble with marriage is that in its extreme interpretation it conflicts with the law of change" (*Hey Rub-a-Dub Dub: A Book of the Mystery and Terror and Wonder of Life,* 217).

13. In *Women and Economics: A Study of the Economic Relation between Men and Women as a Factor in Social Evolution,* 209, 26, 94, 25, 26, Charlotte Perkins Gilman also treats sexual morality within an evolutionary context; Gilman insists that "our belief that a thing is 'natural' does not prove that it is right," and affirms that "our moral concepts rest primarily on facts." But Gilman critiques "promiscuous and temporary sex-relations," the "wrong" of which "is sociological before it is legal or moral." She affirms monogamous marriage as "orderly," "pleasant," and beneficial to humans. Thus while Gilman deploys an evolutionary framework to support the status quo in sexual mores, Dreiser uses the same tool to subvert it.

Like Dreiser, who maintains "the very problem of knowing anything," Witla affirms "that life was somehow bigger and subtler and darker than any given theory" (*NL*, 221; *G*, 118). Even Angela recognizes that "Eugene, with his superiority, or non-understanding, or indifference to conventional theories," "was not like other men—she could see that. He was superior to them" (*G*, 128). In other words, "superior" people use the idea of uncertainty gleaned from Darwin to undermine the pieties of their culture.

Veblen's persona as detached scholar and scientist and Dreiser's as misunderstood artist both rely on the authority of Darwinian science to underwrite cultural criticism. Like Dreiser's attack on conventions, Veblen contrasts an anachronistic notion of truth (absolute, rigid, normalized) with his modernized version (indeterminate, changing, actual) to criticize existing institutions. While Veblen never suggests that humans can live in a world free of institutions, he assumes that institutions will be inept, if not disabling. Because "[i]nstitutions are the products of the past process, . . . [they] are therefore never in full accord with the requirements of the present" (*TLC*, 191). This style of evolutionary reasoning distinguishes Veblen from the Social Darwinists and their naive optimism.

The evolutionary idea also underwrites one of Veblen's satiric gems, which, in turn, forms a fulcrum of his cultural criticism. Revising Alexander Pope's dictum that " 'Whatever is, is right,' " Veblen's belief in institutional lag causes him to insist that " 'Whatever is, is wrong' " (*TLC*, 207). He repeatedly unmasks as illusory the permanence "imputed to" institutions, and argues that while the "institutions in question are no doubt good for their purpose as institutions, . . . they are not good as premises for a scientific inquiry" (*POS*, 239–40). Dreiser's Eugene Witla could perhaps himself be seduced by this logic.

The authorities whose versions of truth Veblen needs to undercut are neither reactionary art critics nor stubborn fiancées but primarily other social scientists. Thus one of his most sustained attacks on institutional fraud focuses on the "[s]elf-contained systems" of orthodox economists (*ECO*, 8). Veblen charges that their "preconceptions" lead them to foregone conclusions (*POS*, 82). The only things visible through the rose-colored lenses of economic orthodoxy, he argues, are normalized and naturalized ideals. The goal of orthodox theory, Veblen protests, is "to explain the facts away: . . . theoretically to neutralize them so that they will not have to appear in the theory" (*POS*, 249).

Since Veblen understands the maxims of neoclassical economics to be ideologically motivated hypotheses, he delights in deconstructing them. For instance, he cites the concepts of " 'ordinary profits' " and " 'adequate demand' " (*ECO*, 106), his crafty quotation marks raising important questions (for why should any profit be considered "ordinary" or any demand "adequate"?). Such orthodox concepts rationalize as normal and natural activities that serve those with power and money. Strategically distinguishing between "reality" and "fact," Veblen contends that what passes for real in neoclassical theory has a "ceremonial or symbolical force only" (*POS*, 90, 125).[14] He believes that belief in the conclusiveness of any theory inevitably causes its promoter to delude both himself and others.

Veblen's rhetoric of confrontation bears comparison with that of the pragmatist philosophers who also affirmed an evolutionary idea of truth. In a 1909 lecture printed in *Popular Science Monthly*, John Dewey explains that *The Origin of Species* inaugurated a new epoch in philosophy, shifting discussion from the assumption of fixity to the ubiquity of change. Similarly, William James, who believed that all experience is "everlastingly in process of mutation," concludes that "both [reality] and the truths men gain about it are everlastingly" in flux; therefore, all "rationalistic" and a priori schemes are inadequate. Charles Saunders Peirce insists that the philosopher give up absolutism and affirm "fallibilism."[15] The pragmatists, unlike Veblen, remained confident that the intellectual honesty they called for in recognizing evolution would ultimately lead to more secure knowledge.

Veblen's adoption of the pragmatic idea of truth as an evolving construct focuses more on dismantling existing authority than on the possibility of attaining epistemological security. He turns authority against itself by demonstrating that the more established any institutionalized version of truth ("Whatever is"), the more "wrong" it is likely to be. Veblen uses the pragmatic idea of truth for his own ends—for instance,

---

14. Veblen attacks ownership in exactly the same way, reasoning that in the natural rights defense of property, "the endeavor is to make facts conform to law, not to make the law or general rule conform to facts" (*TBE*, 319). Veblen's critique of orthodox economics also participates in what Morton White, in *Social Thought in America: The Revolt against Formalism,* identifies as the "revolt against formalism"—a philosophical challenge to the empiricism of Bentham and Mill for not being sufficiently empirical.

15. John Dewey, "The Influence of Darwinism on Philosophy"; William James, *Pragmatism,* 99 (on "rationalism" see lec. 1); Charles S. Peirce, "The Scientific Attitude and Fallibilism."

when he questions whether the emperor of neoclassical economics, Adam Smith, is really wearing any clothes. According to Veblen, Smith "spoke the language of what was to him the historical present," but that language was already obsolete since it derived from "the recent past of his time" (*AO,* 57).¹⁶ Thus Smith's theoretical constructs could not purport to describe accurately economic activity in eighteenth-century England, much less in twentieth-century America.

Veblen does not limit this point about the obsolescence of theories to Smith; he claims that *all* theoretical constructs are the products of specific historical conditions, and thus inevitably "imbecile" (*IOW,* 25). When one considers Veblen's generalization that the "canons of validity are made for [any thinker] by the cultural situation" (*POS,* 17), it becomes clear that he pioneered a trend that has lately taken a tenacious hold on many branches of cultural studies: insistence upon the "socially produced dimension of our habits of thinking," in the words of a recent commentator.¹⁷ Veblen submitted his own ideas to the caveat that all theories emerge from an evolving social environment and therefore must continuously be revised. As he remarks, "like other men of earlier generations, [the] incoming generation of economists are due to be the creatures of that heredity and environment out of which they emerge" (*ECO,* 3). That sentiment marks what may be Veblen's closest approach to postmodernism.

It would be difficult to establish the validity of one's cultural criticism simply by rejecting existing truth claims, and neither Dreiser nor Veblen attempt to do so. After discrediting static truths with evolutionary logic, they legitimize their projects by aligning them with scientific inquiry. Observation and analysis of empirical "fact" become crucial, although justifying access to facts is problematic, especially facts that have been invisible to other observers.¹⁸ As is no doubt already clear, I do not wish to join the already crowded group of discussants who fault earlier thinkers for their naive faith in science.¹⁹ Dreiser's and

---

16. Veblen again discredits Smith on the grounds of his "historical present" in fact being "the recent past" (*VI,* 27).
17. John Fiske, "Cultural Studies and the Culture of Everyday Life," 164.
18. On the elusiveness of facts, see Thomas Haskell, *The Emergence of Professional Social Science: The American Social Science Association and the Nineteenth-Century Crisis of Authority,* 104.
19. One of the most influential examples of critical condescension toward Dreiser for believing in facts comes from Lionel Trilling in *The Liberal Imagination.* In *All Is*

Veblen's works share a commitment to empirical science, but rather than considering the philosophical validity of their position, I am focusing on its instrumentality in their cultural criticism.

But first, a disclaimer about Dreiser's and Veblen's attitudes toward, and uses of, scientific fact may help to forestall misunderstanding. Rather than the positivistic assumption that science can lead to absolute knowledge, they assume, as Veblen puts it, that "the outcome of any serious research can only be to make two questions grow where one question grew before"; thus, approaches that question existing theoretical schemes—such as Veblen's and Dreiser's—are more reliable than any received answers (*POS*, 33). Dreiser and Veblen use science not to locate a value-free space for inquiry but to authorize cultural criticism. They thus cash in on the prevailing scientism of their era while negotiating a way around the alleged value-neutrality of positivism.[20]

In 1907 William James remarked, "Never were as many men of a decidedly empiricist proclivity in existence as there are at the present day. Our children, one might say, are almost born scientific."[21] Veblen explains this appeal of science as largely a matter of methodology. A scientific conclusion remains "decisive so long as it is not set aside by a still more searching scientific inquiry" (*POS*, 4). In other words, modern scientific methodology carries such authority because of its provisional and experimental nature. The pursuit of "facts," not "putative phenomena warily led out from a primordial metaphysical postulate," also permits the modern scientist to proceed "by observation rather than by ratiocination"—by looking rather than theorizing (*ECO*, 150).

True: The Claims and Strategies of Realist Fiction, Lilian R. Furst shows how discussion of literary realism has been stymied by critics' assumptions that referentiality (the novel as portraying an external reality) and textuality (the novel as a purely verbal product) are mutually exclusive. Examining "the credibility of the illusion conjured up in the realist text" (16), Furst shows an admirable tolerance for paradox. See also Susan Rubin Suleiman, *Authoritarian Fictions: The Ideological Novel as a Literary Genre,* which takes up French *romans à thèse*, defined as a subdivision of the realist novel that flourishes at times of social or ideological conflict; and Barbara Foley, *Telling the Truth: The Theory and Practice of Documentary Fiction.* For an interdisciplinary treatment of the phenomenon of realism, see David E. Shi, *Facing Facts: Realism in American Thought and Culture, 1850-1920.*

20. In *Business Enterprise,* Veblen declares positivism dead (367 n 1 and 371). In *The Origins of American Social Science,* 143, Dorothy Ross describes Veblen as a positivist but an unusual one: "the great exception to . . . liberal revisionism, for he retained his Gilded Age socialism, and claimed for it the warrant not of ethics, but of positivist science."

21. James, *Pragmatism,* 10.

Dreiser, with "his insatiable lust for facts" and "fixation on details," also affirms the preeminence of observation. For instance, in *A Book about Myself,* he proudly records the day a copyreader for the *Chicago Daily Globe* told him, "You have your faults, but you do know how to observe.... Maybe you're cut out to be a writer after all, not just an ordinary newspaper man" (*BM,* 67). Numerous scientific observations crop up in his novels to explain human behavior, from the anastates and katastates in *Sister Carrie,* to the lobster, squid, and mycteroperca bonaci of *The Financier* (1912), to the Freudian language of *An American Tragedy* (1925). Such references point to the ancestral prominence of Emile Zola in the genealogy of Dreiser's thinking. In 1880 Zola declared that "the experimental novelist is nothing but a special kind of scientist, who uses the tools of other scientists, observation and analysis." Fifty-five years later, Max Eastman (who commends both Dreiser and Veblen in *The Literary Mind: Its Place in an Age of Science* [1935]), extended Zola's scientific analogy to declare the novel "a species most admirably adapted for survival in this practically scientific world."[22] But while Dreiser's scientific posturing can be traced back through his pedigree as a naturalistic novelist, it also crucially informs his self-perception as a cultural critic. He consistently invokes observation and empiricism to discredit traditional morality, which he simply charges with not "accord[ing] with the facts of life as I have noted or experienced them" (*HRDD,* 252).

Two essays that manipulate social scientific discourse, Dreiser's "Neurotic America and the Sex Impulse" (1920) and Veblen's "Dementia Praecox" (1922), show the instrumentality of science in their cultural criticism. These essays marshal new theories of psychology and lean on loose applications of Darwinian theory to indict postwar America for adolescent hysteria.

The rhetorical strategy of "Neurotic America" sets up America as the case study, Dreiser as the sage psychoanalyst. Within the first paragraph he places his observation on the side of the "calm and exhaustive study" capable of "scientific conclusion" and advances authoritative phrases such as "deep-seated neurosis" and "psycho-analysis" (*HRDD,* 126).

---

22. H. L. Mencken, "Dreiser's Novel the Story of a Financier Who Loved a Beauty," 747; Kaplan, *Social Construction,* 152; Emile Zola, "The Experimental Novel," 193; Max Eastman, *The Literary Mind: Its Place in an Age of Science,* 225. For a wideranging analysis of what observation meant to Dreiser, see Carol Shloss, *In Visible Light: Photography and the American Writer, 1840-1940,* chap. 3.

Over the course of the essay Dreiser drops the names Krafft-Ebing, Ellis, and Freud. His point (which still rings true) is that Americans typically insist on sexual virtue, while hypocritically and eagerly consuming accounts of sexual vice: films with sultry movie goddesses, soft porn, and especially, accounts of sex crimes all play to packed audiences. Dr. Dreiser "psychoanalytically trace[s]" this contradiction between what Americans profess and what they desire to diagnose "strangely exaggerated (neurotic, I think) conceptions of the part sex or its overemphasis plays in life due to repression, which have followed upon impossible religious theories brought from abroad" (*HRDD*, 131). So Dreiser's criticism of American sexual hypocrisy slips in the back door opened so authoritatively by psychoanalysis.

Dreiser's other subjects in "Neurotic America," an attack on censorship and a defense of sexual varietism, emerge from behind the pseudoscientific observation on American hypocrisy. He has considerable personal investment in these subjects and handles them in ways familiar to readers of Veblen. Dreiser redefines the aberrant as the norm: "the impulses we are trying to suppress are . . . perfectly normal. . . . [But w]e are afraid to face ourself honestly and openly" (*HRDD*, 131). He also characterizes himself as seeing the invisible: "scarcely one of a hundred American men or women view this phase of life intelligibly. . . . Usually before their so-called vision . . . hangs an inane and miasmatic cloud of cant and make-believe" (*HRDD*, 135). In "Neurotic America," the prestige of psychology initially legitimizes Dreiser's cultural criticism, and he also frames his argument within a loosely Darwinian idea that the nation has far to go on its evolutionary path: "the thing we think we want is an infantile conception of life and its processes, unsuited to thinking men and women" (*HRDD*, 131).

One of Veblen's most fascinating rhetorical performances makes a similar point about United States culture and deploys the same logic. Like "Neurotic America," "Dementia Praecox" examines censorship and argues that one can correlate the evolution of a society with the mental stability of its members. In Veblen's version of the national case study, the postwar "situation in America is by way of being something of a psychiatrical clinic. . . . [T]he case of America is . . . a certain prevalent unbalance and derangement of mentality" (*ECO*, 429). Veblen, pretending to be a psychiatrist, remarks that dementia praecox (i.e., precocious dementia) is most common among adolescent males but that no cohort is exempt from contamination. The symptoms include "fearsome and feverish credulity," paranoid panics over "footless outrages and odious plots and machinations," "intolerance," incoherence, superstition, and, he slips in, increased church attendance (*ECO,* 429-30). As with "Neurotic America," it is not difficult to locate embedded within "Dementia Praecox" a spirited critique of America's entrance into World War I and the resulting hysteria and paranoia, at home and abroad.

These essays make clear a tendency visible in many of Veblen's and Dreiser's works: claims for scientific detachment accompany stinging indictments of the status quo. The prestige of science justifies supposedly objective attacks that are actually belligerent. This reliance on science no doubt looks dubious today. As Evelyn Fox Keller remarks, "The objectivist illusion" and "investment in impersonality . . . constitutes the special arrogance . . . of modern man, and at the same time reveals his peculiar subjectivity." Although feminist and other poststructural critiques of scientific objectivity make this aspect of Dreiser's and Veblen's cultural criticism appear dated, we must recall that, at the time, the prestige of science made the strategy convincing and authoritative. The appeal of objectivity for Veblen and Dreiser's generation can perhaps best be grasped by considering Mary Douglas's claim that "[t]he striving for objectivity is precisely an attempt not to allow socially inspired classifications to overwhelm the inquiry."[23] Dreiser and Veblen hoped to use science to get past the existing institutional frameworks they believed distorted reality.

Empiricism against Morality

In his study of the founding of the United States historical profession, Peter Novick comments that "the objectivity question is far from being 'merely' a philosophical question. It is an enormously charged emotional issue: one in which the stakes are very high, much higher than

---

23. Evelyn Fox Keller, *Reflections on Gender and Science,* 70; Douglas, *How Institutions Think,* 59. In "Naturalism in American Literature," Malcolm Cowley identifies what I am describing as Dreiser's (and Veblen's) emotional investment in objectivity as characteristic of literary naturalists: "Their objective point of view toward their material was sometimes a pretense that deceived themselves before it deceived others" (325). In "Thorstein Bunde Veblen, 1857-1929," John Maurice Clark remarks on "one of the unanswered puzzles" in Veblen: "his own attitude toward this subjective element entering into his avowedly objective treatment" (598).

in any disputes over substantive interpretations." As seen in "Dementia Praecox" and "Neurotic America and the Sex Impulse," Veblen's and Dreiser's claims for scientific objectivity can be likewise emotionally charged. They consistently use empirical data to contest a range of moral authorities, but their most common target is Christianity. Their strategies for assailing the church reveal how they pretend scientific detachment to pass off volatile, emotional interpretations. Their attacks on Christianity also provide a striking (and strikingly literal) confirmation of Edward Said's claim that the modern intellectual stands in necessary opposition to sacred views of the world.[24]

Two of the most provocative treatments of Veblen describe him as a moralist, but little attention has been paid to the extent that analysis of religious authority saturates his writing.[25] Veblen, who considers religious and scientific authorities fundamentally incompatible, once remarked that a "devout scientist" was as rare as a "white blackbird" (*HLA,* 150). Good religion makes poor science: "[I]n the institutional scheme of the civilised nations the beginning of wisdom is the fear of God; whereas in the technology of physics and chemistry the beginning of wisdom is to forget Him" (*AO,* 282). Scientists have taken over some of the ground once held by priests and prophets, "gradually encroach[ing] on the domain of authentic theory previously held by other, higher, nobler, more profound, more spiritual, more intangible conceptions . . . of knowledge" (*POS,* 50). Christianity's lost authority, in other words, is the scientist's gain.

Veblen uses the opposition of science and religion to challenge authority invested in both persons and institutions. For instance, reviewing John Bates Clark's *The Essentials of Economic Theory,* he invalidates the ideas of his Carleton College professor. Clark cannot, reasons Veblen, write a valid economics text because he is "a spokesman for the competitive system" (*POS,* 189). Although the partisanship Veblen has in mind is largely political, a footnote to his critique of Clark posits that moral partisanship can likewise corrupt scientific inquiry:

24. Peter Novick, *That Noble Dream: The "Objectivity Question" and the American Historical Profession,* 12; Said, *Representations,* 88–89, chap. 6.
25. See Aaron, *Men of Good Hope,* chap. 7, and David Noble, "The Sacred and the Profane: The Theology of Thorstein Veblen." Most mentions of Veblen and religion are purely metaphoric: for instance, Max Lerner's comment in his introduction to *The Portable Veblen* that Veblen became "the terror of received truth in economics as Luther had once been the terror of received truth in religion" (19).

> What would be the scientific rating of the work of a botanist who should spend his energy in devising ways and means to neutralize the ecological variability of plants, or of a physiologist who conceived it the end of his scientific endeavors to . . . denounce and penalize the imitative coloring of the Viceroy butterfly? What scientific interest would attach to the matter if Mr. Loeb, e.g., should devote a few score pages to canvassing the moral responsibilities incurred by him in his parental relation to his parthenogenetically developed sea-urchin eggs? (*POS*, 189 n)

Veblen's satirical questions assume the primacy of science in order to invalidate as moral meddling any sort of ideological commitment on the part of an investigator.

His contempt for moralism leads Veblen to contrast what he calls "scientific" with "homiletical" uses of data: science pursues "facts" and is trustworthy; homily assumes "faith" and is epistemologically unsound (*POS*, 305). He finds economic theory especially prone to contaminate its scientific basis by moralizing. Whether describing the self-fulfilling nature of John Bates Clark's theories as "afford[ing] a basis for those who believe in the old order . . . , and to confirm in the faith those who already believe in the old order," or Adam Smith's legacy as "devout optimism," Veblen invalidates orthodox economic theory by locating traces of religion in it (*POS*, 206, 118).

Veblen's assault on the premises of neoclassical economics proceeds like an attack on theodicy. In fact he claims that most social scientists preoccupy themselves with devising "taxonom[ies] of credenda" (*POS*, 21). The problem with these credenda (or articles of faith) is that when economists take as "absolute truth" one of their own fictitious concepts, such as natural law, or treat an institution, such as invested wealth, as if it were "sacred," a "constraining normality . . . of a spiritual kind" keeps them from accurately observing the economy (*POS*, 61, 358 n, 62). Economic constructs that serve business interests function like sacred dogmas in that people worship them, fear to question them, and assume—like devout commentators on natural disasters—that they serve the greater good in mysterious ways. No wonder that economists who defend business enterprise show the same sort of "reasonably intolerant temper" that characterizes the "devout intolerance of the Church" (*ECO*, 12; *AO*, 47).

Veblen links received economic theory with Christianity because he believes both institutions support a scandalous inequality of power.

Behind the outrageous satire of his best-known commentary on Christianity, in chapter 12 of *Leisure Class,* lies a starkly serious point. Noting the "elaborate system of status" revealed by the "superhuman vicarious leisure class of saints, angels, etc.," Veblen sees "the archaic habitual sense of personal status,—the relation of mastery and subservience" (*TLC,* 317, 331). He draws attention to "the term 'service' " as applied to religious rituals in *Leisure Class* and, returning to a cognate word in another essay, notes that " 'Servant' implies 'Master,' of course" (*TLC,* 122; *ECO,* 270). He further discloses the oppressive uses of Christianity in a cutting essay on "Christian Morals and the Competitive System" (1910). It was in the beginning, is now, and ever will be, says Veblen, primarily the lower classes who practice the distinctively Christian habit of nonresistance, and so the church contributes to the maintenance of an invidious status system (*ECO,* 207-9). Veblen's terms for this master-slave paradigm, which he believes Christianity along with orthodox economic theory perpetuates, are the "Vested Interests" and the "Common Man."[26]

In an endnote to chapter 11 of *Absentee Ownership,* Veblen aligns Christianity with capitalism through some of his most devastating satire. Published two years before Bruce Barton celebrated Jesus Christ as the ancestor of modern American businessmen in *The Man Nobody Knows,* four years before Sinclair Lewis's *Elmer Gantry* exposed the salesmanship of salvation, Veblen's *Absentee Ownership* showed "how truly business-as-usual articulates with the business of the Kingdom of Heaven" (320). He characterizes Christianity as a "fabrication of vendible imponderables in the nth dimension" and recommends that secular salesmen take note of how Christians have perfected the businesslike art of promising everything—even salvation—and delivering nothing (*AO,* 325, 320).

Veblen's seven-page endnote, complete with two footnotes of its own, also draws attention to another sort of authority. Like Alexander Pope in *The Dunciad* and Jonathan Swift's satire on brainy but impractical Laputans, Veblen plays serious games with scholarly pretension. Notably, he attacks authority most vitriolically in footnotes, as seen in this damnation of salvation and in the comment about Viceroy butterflies and Loeb's sea urchins cited earlier. Veblen's footnotes display his mastery of and his contempt for scholarly writing techniques, while

26. On this important distinction see especially *Vested Interests.*

cordoning off especially polemical material by excluding it from the text proper.[27] Another typically Veblenian use of footnotes—to cite his own earlier works—also plays consequential games with scholarly authority, reminding readers of the self-interest and vanities of scholarship.

As one of Veblen's classic rhetorical ploys uses footnotes to partition off volatile material from the text proper, he similarly justifies his attacks on Christianity by a startling exclusion: he refuses to consider moral questions. In his attack on the salesmanship of salvation, for instance, Veblen will examine only the "workday factors" of religion, and these in a "detached and objective" manner, "leaving all due sanctimony on one side for the time being, without thereby questioning the need and merit of such sanctimony as an ordinary means of grace" (*AO,* 321). And in "Christian Morals and the Competitive System," Veblen explains he can entertain "no question . . . as to the intrinsic merit, the eternal validity, of either" (*ECO,* 202). Perhaps most outrageously, Veblen's analysis of the church's attitude toward illegitimate births (an issue of special concern to Dreiser) posits that the idea of sin evolved out of the economic need of the clergy: since babies born out of wedlock do not produce the "sacramental excise from which the clerics of Holy Church derive a substantial portion of their livelihood," bastard children "have had the fortune to become sinful" (*ECO,* 241 n).

Veblen can exclude morality from discussions of religion only because he poses as a scientist impartially examining data. He claims to observe only the institutional aspects of religion and the habits of thought they foster. He even audaciously denies that the words *wrong* or *right* convey any moral judgment: "They are applied simply from the (morally colourless) evolutionary standpoint" (*TLC,* 207). Conveniently remembering during one of his diatribes that he is an economist, Veblen points out that "[t]he moral, as well as the devotional value of the life of faith lies outside of the scope of the present inquiry" (*TLC,* 293).

---

27. In another footnote, Veblen cites the *Report of the Industrial Commission,* vol. 1, to support his audacious declaration that "the heroic rôle of the captain of industry is that of a deliverer from an excess of business management. It is a casting out of business men by the chief of business men" (see *TBE,* 48-49 and n 1). Nowhere is Aaron's comment (*Men of Good Hope,* 217) that Veblen *uses* the form of the academic monograph for his own ends better illustrated than in his outrageous footnotes. For an analysis that is Veblenian in spirit, see Joseph Bensman, "The Aesthetics and Politics of Footnoting."

In a list of socially repressive forces that was published in 1930, Dreiser includes "Religion as opposed to science" (*SUP,* 259). Like Veblen, he charges the church with blockading the path to scientific truth and defines his own authority against that of Christianity. Dreiser's characteristic strategy for contesting religious authority, however, differs from Veblen's: he would confront rather than exclude the moral ground. "Shut up the churches, knock down the steeples!" he thunders in *A Hoosier Holiday* (1916; 183).[28] An even more pointed difference is that whereas Veblen's rhetoric of confrontation pleads his detachment, Dreiser's insists on personal engagement. In contrast to Veblen's artful dodging around moral questions, Dreiser uses the confessional mode, thus paradoxically enlisting subjective experiences to make the case for his objectivity.

In openly drawing so many of his facts from personal experience, Dreiser anticipates the recent autobiographical turn in literary and cultural analysis. Dreiser uses his personal experience to challenge institutionalized power. As if to make himself a one-man rebuttal of Catholic doctrine, Dreiser repeatedly mentions the sexual irregularities of his sisters' lives, the crippling effect of his father's religiosity, the repressiveness of Catholic schools, and his own youthful fears of sexual sins. Dreiser describes his staunchly Catholic father wasting his entire life thinking about an "entirely mythical heaven" where he hoped to become "a standardized angel—wings, harp, robes and all" (*HRDD,* 112). Although in novels such as *The Bulwark* (1946) and *Jennie Gerhardt* (1911) Dreiser seems more tolerant of religious dispositions such as his father's, the autobiographies typically present John Paul Dreiser as "fanatical," "mentally a little weak," and, most damaging, compare him to a drunkard or drug addict, "addicted to religious formulæ, and with equally fatal results" (*D,* 4, 6, 349).

Dreiser also uses personal data to rebut Christianity in the many graphic descriptions of his sexual experiences. He describes his amorous adventures not only to illustrate his narrow escape from the moralism that he believes ruined his father, but also, and quite strikingly, on the ground of furthering objective truth. In *Dawn,* for

28. Zanine in *Mechanism and Mysticism* and, more emphatically, Lawrence E. Hussman Jr. in *Dreiser and His Fiction: A Twentieth-Century Quest* argue that Dreiser eventually adopts a philosophy of spiritual harmony. Nevertheless, that resolution was a long time coming, and most of Dreiser's books—fiction and nonfiction—portray little tolerance for organized religion.

instance, Dreiser describes when he began masturbating and the ostensibly ensuing physical and mental side effects, including pimples and depression. It is as if the mature Dreiser were diagnosing not only his youthful self but also a puritanical body politic, for he claims to talk so frankly about masturbation in order to disprove the moralists' claim that degeneration inevitably follows so vile a sin (*D*, 268–72). "Poor, ignorant humanity!" writes Dreiser, offering up his own experience to educate others: "I wish that all of the religious and moral piffle and nonsense from which I suffered in connection with this matter [masturbation] could be undone completely for the rest of the world by merely writing about it" (*D*, 272). In a strategy typical of his autobiographies, Dreiser draws upon the authority of personal experience to achieve an air of superiority over the insecurities he felt as a youth (and which, he implies, many adult Americans still feel). He thereby also positions himself as superior to the traditional moral values he critiques and encourages readers to draw on the authority of their own experience to free themselves.

Had Dreiser's autobiographies been published as he first wrote them, they would have been even more sexually explicit. In the Pennsylvania edition of *Newspaper Days* (1991), which restores material excised from the 1922 edition, Dreiser recounts his first experience of fellatio. Although this frank passage begs for a variety of interpretations, I would like to draw attention to Dreiser's suggestion that the guilt induced by Catholic doctrine promotes a truly perverse sexual pleasure:

> I felt as though I were witnessing one of the great horrors and crimes of the soul and the world, a thing, or sin, which once one passed out of life, might certainly cause one to be grilled in hell, and that rightly, as the church had taught. And yet, in spite of my great shame and horror, or perhaps all the more because of it, it was all too delicious. . . . Life or God might punish me for this! Maybe there was something to the dread Catholic doctrine that this was mortal sin, punishable by eternal fire! (*ND*, 593)

To his credit, Dreiser is fully aware of the hypocrisy of this prurient spectatorship of his own "delicious" damnation. Indeed, that is the point. He implies that the perversion is neither fellatio nor prostitution, but the Catholic doctrine that made him identify those practices as sinful— and therefore become all the more aroused. Dreiser then condemns his youthful self, especially for his subsequent condescension toward the prostitute, as a "snooping, fearsome, moralistic ass" (*ND,* 594). This passage provides a paradigmatic instance of Dreiser making use of his experiences to discredit conventional morality. Despite the personal nature of the account, Dreiser purports to scientific disinterestedness.

Such moments, so typical in Dreiser's work, illustrate Foucault's claim that "Western man has become a confessing animal." Recently, Irene Gammel has adopted a Foucaultian perspective and cast a skeptical eye on Dreiser's talk about "liberating" sexuality and his equating truth with writing frankly about sex. Gammel uses Foucault to discredit Dreiser's belief in a natural sexuality that needs only to be liberated from the shackles of society; in particular, her analysis of "Emanuela" in *A Gallery of Women* shows Dreiser, far from liberatory in his attitudes toward sexuality, participating in the hysterization of women's bodies.[29] Nevertheless, while Dreiser's liberatory rhetoric may indeed be suspect, and while it has become axiomatic to say that sexuality is always constituted, never natural, Dreiser did *re*constitute sexuality discursively, challenge assumptions about sexual appropriateness (in literature and in life), and critically revise the traditional "fallen woman" plot to show female characters succeeding rather than failing (Carrie Meeber) or, no less significantly, living while their lovers die (Jennie Gerhardt). And while Dreiser does seem to hold "natural," untrammeled sexual expression as an ideal, he also understands sexuality to be produced by specific historical conditions (as seen especially in the polemical passages of *The "Genius"*). To say that Dreiser simply uses sexuality to reinscribe his own power locks us into a pattern of infinite regress that tends to neglect the literary-historical changes he inaugurated.

A more favorable reading of Dreiser's treatment of sexuality can be constructed from the same Foucaultian vantage point. Foucault's revolutionary claim that sexuality, far from being repressed beginning in the seventeenth century, became the subject of incessant discussion, rests on the pervasiveness of the confessional mode. The confession, derived from Catholicism and refined by pedagogical, medical, and (especially) psychoanalytic imperatives to "confess" the truth about one's sex, comprises an "agency of domination" in which the confessor

29. Michel Foucault, *The History of Sexuality,* vol. 1: *An Introduction,* 59; Irene Gammel, *Sexualizing Power in Naturalism: Theodore Dreiser and Frederick Philip Grove,* chap. 5.

submits to authorities who must interpret the meaning of his sexual thoughts, desires, and actions.[30] But if we examine the effects of Dreiser's confessions of his (and his fictional characters') sexual experiences, we find that they do not confer power on an authority to whom the confession is made, and who must interpret the "truth" of the confession (doctor, priest, educator, etc.). Rather, this lapsed Catholic's confessions assert the authority of the confessor and the integrity of his or her sexual experiences, while challenging the right of anyone else to moralize or condemn. Furthermore, Foucault claims that power in modern societies invests itself less in the enforcement of laws than in the policing of norms, and that sexual confessions reinscribe those norms. Dreiser understands that process and, far from endorsing it, he critiques it in *An American Tragedy,* when Clyde Griffiths's forced confession to a court and angry citizens of his sexual relations with Roberta Alden reveals communal norms winning out over the "truth" of what happened—"truth" that becomes increasingly unclear as various legal and moral authorities attempt to define it. One need not buy into Dreiser's rhetoric of sexual liberation to give him credit for comprehending, and fighting against, the politics of sexuality.

Dreiser's rhetoric of engagement will not "hesitat[e] in revealing the net of flesh and emotion and human relationship" (*D,* 3). Veblen's rhetoric of detachment, however, necessitates that the cultural critic "eliminate the 'personal equation' " (*AO,* 262). Yet Dreiser's self-exposures serve the same purpose as Veblen's hiding and dodging: both use empirical "fact" to underwrite cultural criticism. Dreiser's method of inserting himself, staking out his own position in reference to the object of his analysis, would horrify Veblen, but it forms an important precursor to the "identity politics" celebrated, reviled, and debated by various camps of cultural critics today.[31]

Besides being derived from personal experience, the facts Dreiser invokes to contest Christianity are less technological and more biological than those Veblen uses. A typically Dreiserian argument contrasts biological needs with religious norms to show that "[n]ature knows better than man what it needs" (*D,* 143). For instance, he describes promiscuity as problematic not for the varietist, as the traditionalist

---

30. Foucault, *History,* pt. 3, 62.
31. In *Essentially Speaking: Feminism, Nature and Difference,* Diana Fuss defines *identity politics* as "the tendency to base one's politics on a sense of personal identity" (97), a definition that clearly applies to Dreiser.

would claim, but for moralists themselves: "Nature seemed to be without Christianity or Christian morals, and this shocked the American terribly" (*HRDD*, 40). Dreiser considers scandalous the pretense of the "non-existence of such things" as prostitution (*ND*, 144). He never seems to tire of disputing received moral explanations: "in spite of all the theories . . . , life . . . appears to be chronically and perhaps incurably varietistic" (*HH*, 366).

Like Veblen's objection to fitting facts into preconceived theories, Dreiser criticizes attempts to force reality into a preconceived system. Thus in *Dreiser Looks at Russia* (1928)—an account of a trip in which Dreiser praises theoretical aspects of Communism while grousing about the particulars of daily life in Russia—Dreiser comes down most harshly on what he defines as "propaganda." He compares Soviet propaganda to American advertising (*DLR*, 89–90) and, even more revealing of his own thought processes, to Catholic dogma. "Could the Catholic Church do better" in brainwashing its constituency? Dreiser asks (*DLR*, 96). Catholicism has "a new and, I think, . . . fatal, competitor" in Communist dogma (*DLR*, 96). That comparison signals the depth of Dreiser's reservations about Communism.

Yet according to Dreiser, American religiosity, with its "peculiar, . . . fierce determination, to make the Ten Commandments work," is the biggest offender (*HRDD*, 262). Dreiser responds by piling masses of data on the stairs of the church. Because received religious schemes accord neither with Dreiser's personal experience nor with what he sees around him, he concludes they must be wrong. He lists numerous newspaper accounts of everyday tragedies "to show how . . . impossible of a fixed explanation or rule" the world is (*HRDD*, 10). Dreiser challenges readers to "locate Divine Mind, Light, Wisdom, Truth, Justice, Mercy" in daily news items (*HRDD*, 13). He feels that Christianity maintains its harmonious theories only by ignoring, as Veblen puts it, all such inconvenient facts. Although "The religionists do not need to explain anything since they ignore facts," Dreiser would use observation and inductive reasoning to discredit dogma. The "great realistic facts" will "contradic[t] all the noble fol-de-rol of the puritans and the religionists" (*ND*, 609; *BM*, 212). Dreiser tellingly explains his motive for writing one of his anti-Catholic essays: "It is only because I cannot understand why people cling so fatuitously [*sic*] to the idea that there is some fixed idyllic scheme or moral order handed down from on high, . . . that I write this" (*HRDD*, 8). Dreiser, in other words, feels he has no axe to grind; he is just setting the record straight.

Dreiser's novels frequently adopt this polemical empiricism to refute moralized readings of the world. In *An American Tragedy*, for instance, the narrator describes Roberta's pregnancy—the decisive empirical fact in a novel filled with ambiguity—as "one of those whirling tempests of fact and reality in which the ordinary charts and compasses of moral measurement were . . . of small use" (388). Facts speak louder than morals in this novel (as in most of Dreiser's works), and it is morals that change with circumstances. As Clyde and Roberta's unsuccessful attempts to secure an abortion reveal, they are held to stiffer moral codes than would pertain for wealthier lovers. That sort of observation led Dorothy Dudley, one of Dreiser's early promoters, to remark that he steadfastly "refus[ed] to celebrate the substitute for the real."[32] Veblen would call this stance the denial of ceremonialism.

Like Veblen's attack on the selling of salvation, Dreiser contemptuously describes the "sacrament and indulgence salesmen" who run Catholic schools (*D*, 26). He questions why "[s]hould only the great church organizations be permitted to monopolize a false and yet so profitable field?" He sees political organizations as having "usurped the religion business, gone, wholesale, into the business of electing officers or heads, building churches, collecting money and using the same . . . to erect and enjoy power" (*D*, 473-74). This power differential is entrenched by the catechism, about which Dreiser remarks, "assure a man that he has a soul and then frighten him with old wives' tales as to what is to become of it afterwards, and you have a hooked fish, a mental slave!" (*D*, 138-39). He asserts that when dogma rules the mind, "the incentive to discover is gone" (*HH*, 463).

Religious education comes in for particular ridicule. Dreiser dismisses without reservation the Catholic school he attended in Evansville, Indiana, for two years where he "never learned anything about anything" (*HH*, 461). Battering Catholic education with facts, Dreiser asks, "Hasn't the world had enough of unsubstantiated dogma by now? Why permit an unwitting child to be dosed with . . . religious and social folderol when there are masses of exact data at hand?" (*D*, 129). Duping the common man by narrowing choices into the oversimplified options of good and evil, the religious educational system resists change and helps maintain the status quo.

Dreiser and Veblen contest religious authority as hostile to reality in general and to their own enterprises in particular. Christianity

---

32. Dorothy Dudley, *Forgotten Frontiers: Dreiser and the Land of the Free*, 38.

exemplifies what Veblen calls an "imbecile institution"; once entrenched, it perpetuates its power so successfully as to seem eternal and unchanging (*IOW,* 25). As Veblen remarks, all institutions give the illusion of being "securely and eternally right and good" during their reign (*VI,* 4). By basing its authority on the idea of the eternal, Christianity capitalizes on the "human propensity to hold fast that which once was good," causing its advocates to deny historical change (*AO,* 6). Thus Veblen's particular opposition to religion: while institutions are "always archaic . . . in some degree," they become more so the further removed they are from quotidian life. Consequently, politics is "always more archaic" than more pragmatic arenas of human behavior, and religion is "even more archaic" than politics (*AO,* 43, 44 n). Dreiser and Veblen both argue that religion serves human need rather than revealing divine will: in Dreiser's words, "Our insistence on the existence of general truths does not make them truth"; in Veblen's, "What is proved by the tenacity with which we cling to our teleological conception of the world is, [only] that the constitution of our intellect demands this conception" (*NL,* 216; *ECO* 186). Dreiser's comment that "there is no system ever established anywhere which is wholly good," like Veblen's "Whatever is, is wrong" (*HH,* 367; *TLC,* 207), points to the inevitable futility of all theoretical systems that purport to permanence. "Life will not be boxed in boxes," claims Dreiser (*HH,* 285). Veblen explains the reason why boxes must be forgone: "In Darwinism there is no such final or perfect term, and no definitive equilibrium" (*POS,* 417).

Romancing the Margins

In the process of attacking powerful institutions, Dreiser and Veblen helped to create an alternative sort of authority for the cultural critic. Perceiving themselves as nonconformists, even as socially marginal, and explaining resistance to their ideas as a function of their intransigence, Dreiser and Veblen developed effective personae as oppositional intellectuals. They drew upon their self-perceptions as outsiders to make a virtue out of the tendency of American culture to marginalize independent thinkers. As saboteurs of the status quo, Dreiser and Veblen placed a premium on iconoclasm and romanticized the virtues of being on the margins.

Werner Sollors has recently remarked that, "[i]n America, casting oneself as an outsider may in fact be considered a dominant cultural trait." That observation cautions a healthy skepticism about such protestations as Dreiser and Veblen make to be outsiders, but it also suggests the appeal and the durability of their critical stance. Indeed, many avant-garde intellectuals have recently rediscovered the virtues of the margins so avidly courted by Veblen and Dreiser. The radical feminist bell hooks, for instance, sees "marginality as much more than a site of deprivation. . . . it is also the site of radical possibility, a space of resistance." Similarly, Giles Gunn remarks that "the essential challenge [for the American cultural critic] is to develop a voice that is attentive, answerable, and, as much as possible, unassimilable." One of the most impassioned of recent calls for intellectuals to actively cultivate marginality comes from Edward Said. In *Representations of the Intellectual,* Said defines the intellectual as a "spirit in opposition . . . against the status quo." Rather than placating his audience, the intellectual needs "publicly to raise embarrassing questions." In fact, argues Said, the true intellectual must be an "exile."[33]

Dreiser and Veblen cultivated such a position of exile. Veblen asserts that it takes an "outsider" to see a culture clearly, and an economist, he says, has no choice but to look "in perspective from the outside" (*HLA,* 2; *AO,* 163). Declaring independence from dominant cultural values, Veblen says, "Any given ground of distinction will seem insubstantial to any one who habitually apprehends the facts . . . from a different point of view." This habitual difference constitutes Veblen's professional signature.

It is probably also the most familiar aspect of the Veblen legend. Scholars have characterized him as "alien twice over," "citizen of nowhere by nature," "visitor from Mars," an "unacclimated alien," and "village iconoclast" and noted his "intellectual orphanhood" and "automatic reaction against anything given." Joseph Dorfman is only the most authoritative of many readers who not only cite, but also endorse, Veblen's legendary aloofness: The " 'man from Mars' vantage point . . . is an especially valuable aid to insight in times of unusually rapid and far-reaching social and economic change. . . . Indeed, it is reasonable to

---

33. Werner Sollors, *Beyond Ethnicity: Consent and Descent in American Culture,* 31; bell hooks, "Marginality as Site of Resistance," 341; Giles Gunn, *Thinking across the American Grain: Ideology, Intellect, and the New Pragmatism,* 18; Said, *Representations,* xvii, 11, xvi. It may be the quest for an "unassimilable" position that leads many cultural critics to court the margins. On the "spatial topography of center and margin on which oppositional criticism subsists," see Henry Louis Gates, "Good-bye Columbus? Notes on the Culture of Criticism," 257-58.

query whether the particular kind of original, fundamental speculations offered us by Veblen would have been possible had he sprung from the mainstream of our society."[34] If the scholarly literature can be trusted, then Veblen almost perfectly exemplifies Said's intellectual-as-exile.

But the scholarship cannot fully be trusted. Newly opened archives (including Dorfman's own papers) have cast doubt on much of the influential thesis concerning Veblen's alleged alienation and its basis in childhood poverty and discrimination due to his Norwegian ancestry. The problem lies not in Dorfman's perceiving Veblen as an exile but in how and why he so perceived him. By locating Veblen's marginality within a narrow (and dubious) biographical framework, Dorfman set the stage for decades of scholarship to treat Veblen's social theory as oddly detached from the society it purports to examine. Dorfman's thesis has thus allowed scores of commentators to write off Veblen's works as curious productions of a "man from Mars." The amount of attention paid in scholarly literature and economics textbooks to anecdotes illustrating Veblen's personal idiosyncrasies is nothing short of scandalous. The net effect of Dorfman's marginalization thesis has been to marginalize Veblen.[35]

I would like to shift the question of Veblen's alienation from biographical accounts of his childhood to rhetorical analysis of his writings. Edward Said's insight is again invaluable: "while it is an *actual* condition, exile is also a *metaphorical* condition." That important distinction helps us to free Veblen's alienation from the purportedly

---

34. Kazin, *On Native Grounds,* 133; Robert Heilbroner, *The Worldly Philosophers,* 197; W. C. Mitchell, "Thorstein Veblen: 1857–1929," 603; Richard Hofstadter, *Social Darwinism in American Thought,* 65; John P. Diggins, *The Bard of Savagery: Thorstein Veblen and Modern Social Theory,* 50; Aaron, *Men of Good Hope,* 210; Carlton C. Qualey, introduction to *Thorstein Veblen: The Carleton College Veblen Seminar Essays,* 6; Joseph Dorfman, "Background of Veblen's Thought," 127–28. Dorothy Ross, who in *Origins,* 204, 154, follows the consensus in declaring Veblen "the only true outsider" offers a more plausible explanation: Veblen "was relatively unique among social scientists in adapting from anthropology . . . its stance of the alien observer."

One of the few serious challenges to this core element of the Veblen legend has come from the Frankfurt intellectuals; see, for instance, Adorno's "Veblen's Attack." In his introduction to the Mentor edition of *Leisure Class,* C. Wright Mills provocatively suggests that "It has been fashionable to sentimentalize Veblen as the most alienated of American intellectuals. . . . To be conspicuously 'alienated' was a kind of success [Veblen] would have scorned most" (xviii).

35. Dorfman emphasizes the Norwegian American family's alleged poverty and unfamiliarity with English. Many commentators have explicitly or implicitly linked Veblen's "alien" status with his being born to immigrant parents. But this portrayal, as noted in note 10 of the introduction, has increasingly fallen under fire.

"actual" (biographical) grounds on which Dorfman placed it. I would like to extend Said's argument about the "metaphorical" condition of the intellectual's alienation to examine Veblen's rhetoric of alienation—for his writings do project the image of himself as a marginal man. Two leitmotifs of Veblen's works convey the impression of the author as an outsider. One is his construction of the myth of what he liked to call idle curiosity, the keynote of disinterested research and hallmark of the scientist-scholar. The second is the construction of a pantheon of personal heroes in advance of their times with whom Veblen covertly aligns himself. Edward Said writes of the "pleasures of exile," reminding us that an intellectual may enjoy cultivating "different arrangements of living and eccentric angles of vision."[36] In his praise of idle curiosity and iconoclastic thinkers, Veblen demonstrates the pleasures of exile and eccentricity.

Veblen defines idle curiosity as that which makes people "want to know things," in contrast with workmanship, which motivates them to make things (*IOW*, 85). Idle curiosity is one of only three "instincts," along with workmanship and the parental bent, that Veblen considers positive attributes of humans. Commentators typically identify these three instincts as aspects of a master pattern subsumed under workmanship and contrasted with ceremonialism or exploit. But in an important sense idle curiosity stands out: while the parental bent and workmanship encourage people to cooperate with each other, idle curiosity isolates the person manifesting it. As Veblen remarks, "Sporadic individuals who are endowed with this supererogatory gift [idle curiosity] largely in excess of the common run . . . are accounted dreamers, or in extreme cases their more sensible neighbours may even rate them as of unsound mind" (*IOW*, 87). The intellectual may look like an alien—even a pariah. Thus idle curiosity marks the posture of the intellectual who stands (inevitably, according to Veblen) in exile.

Some of Veblen's comments about the isolation of the intellectual do have an autobiographical dimension, although they point to his unsuccessful attempts to establish himself in the academy, not to his childhood. In *The Higher Learning in America* (1918), Veblen alludes to the blacklisting that he believed sabotaged his career; he also remarks that scholars with "too pronounced an addiction to scientific or scholarly pursuits" are unlikely to be awarded tenure (*HLA*, 179,

---

36. Said, *Representations*, 52, 59.

163). That a scholar of Veblen's distinction never received tenure at any institution, and in fact was driven out from several universities on trumped-up charges, forms one of the more embarrassing chapters in the annals of the modern university.

Nevertheless, Veblen's discussion of idle curiosity has important ramifications for the present debates over university tenure and the wider issue of academic freedom. He argues that idle curiosity, or "intellectual initiative . . . [cannot] be reduced to any known terms of subordination, obedience, or authoritative direction. . . . A free hand is the first and abiding requisite of scholarly and scientific work" (*HLA*, 86-87). Veblen published these words three years after the American Association of University Professors was formed to protect the "gray area of academic freedom."[37] If Veblen is correct about intellectual initiative, then scholars *should* be rebellious, and an institution such as tenure is necessary to protect them. The insubordination that Veblen associates with idle curiosity accounts for the current criticism of universities as bastions for "tenured radicals."

But Veblen offers a defense of intellectual freedom that could be extended to give pause to even the most virulent critics of tenure. He argues that such freedom is good for society at large. Echoing and extending upon the 1915 Report of the Committee on Academic Freedom and Tenure of the AAUP, which claimed that intellectual progress results from "complete and unlimited freedom to pursue inquiry and publish its results," Veblen defends idle curiosity as the force driving all of technology forward, thus improving the conditions of human life. "The impressive mutations in the development of thought" that characterize idle curiosity will only emerge, according to Veblen, from a freely ranging mind (*POS,* 43). Thus the intellectual's insubordination is both necessary for the scholar and good for society. Like what is defined as basic research, idle curiosity is not concerned with how knowledge will be applied, yet "counts up finally . . . into the most substantial cultural achievement of the race" (*IOW,* 87). In fact, the autonomy of idle curiosity ensures its social value, and so the intellectual's exile turns out to form the basis of his or her beneficial connection with society. (Veblen, to be sure, did not put it in these terms.) When Veblen explains how protecting individuals who manifest idle curiosity will benefit all of

---

37. Burton J. Bledstein, *The Culture of Professionalism: The Middle Class and the Development of Higher Education in America,* 305.

society, his explanation has special resonance for physical, natural, and social scientists. But Veblen also describes idle curiosity as a "creative factor" related to the impulse to make myths and other narratives, and so protecting intellectual freedom in the humanities and arts also becomes necessary for society as a whole (*IOW,* 85; *POS,* 17).[38]

To diagnose what ails the university, Veblen turns away from the idle curiosity of its faculty and toward the administration of academia. *The Higher Learning in America* (1918), which Veblen wanted to subtitle "A Study in Total Depravity," is an early example of a currently popular genre: the insider's expose of academia. This genre often entails a complex form of cultural criticism, insofar as the subject is often also the agent of criticism. Of all Veblen's books, *Higher Learning* may most readily display present relevance. As might be suspected, Veblen goes after business schools (a degree from which "thereby widen[s] the candidate's field of ignorance" [*HLA* 207 n]), fraternities ("competitive organizations for the elaboration of the puerile irregularities of adolescence" [*HLA* 123]), and college athletics (games of "sentimental rivalry" revealing the "lower motive of unreflecting clannishness" [*HLA* 235]).

But those are all minor targets. Veblen's real anger is directed at the university's impulse toward quantification, which he reads as demonstrating how businessmen and pecuniary ideals have infiltrated academia, and on this theme he shows astounding prescience. As business concerns "infect the university" (*HLA,* 62), administration increasingly takes precedence over both teaching and scholarship. While budgets shrink, the university's chief business becomes chasing after wealthy alumni and faculty celebrities. Veblen argues that the "captains of erudition" who turn universities into "corporations of learning" destroy higher education because the "utilitarian outcome" sought by business contradicts the disinterest of idle curiosity (*HLA,* 85, 30).

Apocryphal tales about Veblen typically (if dubiously) bill him as an indifferent educator, but in *Higher Learning* he expresses concern over what the commercialization of academia costs students. The pecuniary mentality causes the "substantial pursuit of knowledge" to become

---

38. The storytelling impulse that Veblen associates with idle curiosity seems to derive from what he describes as the mind's need for order: "the requirement of totality" (see *ECO,* especially 191-92). In "Veblen on Scientific Creativity," Alan W. Dyer clarifies the creative dimensions of idle curiosity and argues that Veblen's concept of idle curiosity derives from Peirce's theory of abductive reasoning.

"subordinate to the present pursuit of credits"—a fact that many faculty advisors today can confirm (*HLA*, 128). In the "comprehensive system of scholastic accountancy," knowledge becomes trivialized into "mechanical units of academic bullion" (*HLA*, 103). Learning, charges Veblen, should not be "a competitive business," but when the "needs of the higher learning" contradict "the demands of business enterprise," the latter wins out (*HLA*, 97, 48).

Higher Learning shows Veblen's response to two historically linked but conflicting trends in American education. In celebrating idle curiosity, Veblen speaks out for the new research universities, such as Johns Hopkins (where he attended graduate school) and the University of Chicago and Stanford University (where he held untenured appointments). But in excoriating the pecuniary sensibility, Veblen draws attention to the funding of such research institutions by capitalists. Before the Civil War, the largest single gift to an American college had been Abbott Lawrence's $50,000 to Harvard, but in the late nineteenth century, Johns Hopkins received $3,500,000 as a single donation, while Stanford counted one $24,000,000 gift from its namesake, and Chicago topped Stanford with $34,000,000 from John D. Rockefeller.[39]

Like *Higher Learning, The Engineers and the Price System* (1921) critiques bloated and counterproductive institutions ruled by predatory business principles and suggests ways to fix them. An inspiration for the technocracy movement and, more widely, for various attempts at social engineering in the United States, *Engineers* provides one of Veblen's most influential yet misunderstood arguments.[40] The book is usually read either as a serious revolutionary tract or as a joke, but it is neither. By considering *Engineers* as a rhetorical performance that exemplifies idle curiosity, both the uncommitted nature of the revolution it reputedly foments and the seriousness of Veblen's rhetorical play become evident.

The argument in *Engineers,* which like *Higher Learning* spotlights the destructive hegemony of the profit system, begins by drawing attention to Veblen's language and that of his opponents. After providing a lengthy, eloquent discussion of the etymology of the word *sabotage*

39. Walter Metzger, *Academic Freedom in the Age of the University,* 134.
40. On the influence of *Engineers and the Price System,* see Jordan, who in *Machine-Age Ideology* traces the "philosophical bases of social engineering" back to pragmatism and "Veblenism" (13). *Engineers* was originally published as a series of essays in *The Dial* during 1919.

(which, in his day, frequently appeared in charges leveled against radical labor organizers such as the Industrial Workers of the World), Veblen adopts his mask of *faux* innocence to sabotage the status quo. Claiming to regret the "sinister meaning" attached to the word by anti-labor commentators, Veblen suggests that "sabotage" more accurately characterizes the passive and subtle actions of business owners than the violent and overt actions of rebellious workers (*EPS*, 38-40). He proceeds to demonstrate "capitalistic sabotage" as the modus operandi of business: "peaceable or surreptitious restriction, delay, withdrawal, or obstruction" to keep profits up (*EPS*, 40). The business owner's "salutary use of sabotage" prohibits overproduction, which would cause prices, and hence profits, to fall (*EPS*, 42).

It is difficult to dispute the continuing relevance of the consequences Veblen sees in the corporate division between engineers and managers, even if one backs away from his radical conclusions in *Engineers*. Veblen formulates the difference as one between engineers (thinker-producers) and businessmen (seller-predators). He argues that while businessmen grow "increasingly out of touch with . . . mechanical technology," they become mere "captains of finance" who only pose as "captains of industry." Lacking a "competent grasp of the industrial arts," businessmen ineffectively control industries they increasingly cannot understand (*EPS*, 63). But because technology is complex and interlocking, soon "even the ordinary modicum of sabotage involved in the conduct of business as usual will bring the whole to a fatal collapse" (*EPS*, 75). Consequently, Veblen claims that the collective good would best be served by pursuing productivity over profit—throwing out the businessmen and giving control of industry to the engineers.

The relevance of *Engineers* to Veblen's notion of idle curiosity becomes clear when we consider the similarity of its argument to that of *Higher Learning*. Both books propose transforming institutions, only to retreat in their concluding pages from their own blueprints for change. As Veblen says toward the end of *Higher Learning*, his "inquiry is nowise concerned to reform" and the projected "heroic remedy" is "bound to fail" (*HLA*, 270, 276). He takes the same evasive stance toward the end of *Engineers:* his purpose has been "to show, in an objective way, that under existing circumstances there need be no fear, and no hope, of an effectual revolutionary overturn in America" along the lines he has just sketched (*EPS*, 132). Veblen is not being disingenuous. The reason for his strange maneuvering is that while idle curiosity sanctions speculation about anything, it also necessitates disavowing reform, for to advocate anything is, according to Veblen, to grow partisan and therefore unscientific. By imagining the unimaginable, both books comprise tributes to, and exemplars of, idle curiosity. Ending *Higher Learning* and *Engineers* by ducking for cover under idle curiosity, Veblen can sabotage capitalistic institutions while denying that he is doing so.

Over the years, complaints have proliferated that Veblen fails to develop constructive plans for social change. That charge spotlights a problematic area in Veblen's advocacy of idle curiosity—for how do critical intellectuals who abjure the idea of praxis justify work that is purely speculative—but it also indicates a misunderstanding of Veblen's agenda. He disagrees with the pragmatist contention that knowledge should be conceived of as a tool. Instrumentalism proceeds from the assumption that knowledge should serve the human community, but Veblen aspires to be a *non*participant observer of community.[41]

Veblen's rhetoric of detachment, which is central to his persona as a cultural critic, emerges again in the pantheon of heroes he constructs. On the rare occasions when he praises another thinker, the self-projection is dazzling. Veblen's heroes manifest incisive, critical intelligence, and, consequently, don't fit in with their culture. He has a particular fondness for intellectuals who were too "scientific" for their times and whose findings, therefore, were rejected by defenders of the status quo. In Veblen's hands, at times they seem less a community of intellectuals than a secret society.

The basic pattern of Veblen's commendation can be seen in his characterization of David Hume. Veblen bills Hume as a consummate outsider, "gifted with an alert, though somewhat histrionic, skepticism touching everything that was well received" (*POS,* 96). And according to Veblen, "skepticism is the beginning of science" (*HLA,* 181). Hume's reputation during his lifetime illustrates Veblen's myth of the skeptical

---

41. Arthur J. Vidich and Joseph Bensman observe in "Participant Observation and the Collection and Interpretation of Data," in *Small Town in Mass Society: Class, Power and Religion in a Rural Community,* 352-53, that "an observer prefers to keep his identity vague; he avoids committing his allegiance—in short, his personality—to segments of the society. . . . He is 'asked' to answer the question, 'Who do you speak for?' and it is an answer to this question which, in the interests of research, he avoids.

"Consequently, the observer remains marginal to the society or organization or segments of them which he studies. . . . Being both a participant and an observer is 'the strategy of having one's cake and eating it too.'"

scientist as misunderstood prophet: although he "out-Britishes the British," "[t]he skeptic of the type of Hume has never been in good repute with those who stand closest to the accepted religious truths" (*POS,* 97, 106). We can see Veblen's notorious difficulties with university administrators reflected in his claims that while Hume's contemporaries might have considered his views strange, *they* rejected him rather than vice versa. Hume was simply in advance of his time, "too modern to be wholly intelligible to those of his contemporaries who are more neatly abreast of their time" (*POS,* 97).

In "The Socialist Economics of Karl Marx and His Followers: I" (1906), Veblen's characterization of his greatest competitor significantly focuses on discrediting hostile critics. The intellectual peons who carp that Marx offers no proof for his labor theory of value fail to see "a self-satisfied superior's playful mystification of those readers (critics) whose limited powers do not enable them to see that his proposition is self-evident" (*POS,* 419–20). The critics judge Marxist theory against the status quo and "piously" (a significant adverb) believe they have refuted it, "whereas, of course, they have for the most part not touched it" (*POS,* 422). Through what he says about Marx, Veblen's own goal becomes clear: confound your critics by staying so far ahead that none can catch you. He describes Marx as a trailblazer so far in advance of his times that one must follow him on his own terms, or not at all: "It is . . . not by an itemised scrutiny of the details of doctrine and by tracing their pedigree . . . that a fair conception of Marx . . . may be reached, but rather by following him from his own point of departure out into the ramifications of his theory" (*POS,* 413). Given his predecessor's genius for forging a science out of social criticism, it is no wonder that Veblen reads Marx the way Veblen himself wishes to be read.

Veblen's most enthusiastic celebration of intellectual outsiders occurs in his remarkable essay opposing Zionism where, as has often been noted, he projects himself unabashedly. "The Intellectual Pre-Eminence of the Jews in Modern Europe" (1919) provides a sincere tribute to Jewish "pathfinders and iconoclasts, in science, scholarship, and institutional change and growth" (*ECO,* 224) while advancing Veblen's own views.

The essay is a classic of Veblenian rhetoric in that the ostensible subject, Zionism, serves as the pretext for a long digression containing the actual agenda: an explanation of why the best scholars and scientists never fit in with their culture. The cardinal virtue of Veblen's intellectual Jews is their homelessness. These "wanderer[s] in the intellectual no-man's land" are superb scholars precisely because of their alienation (*ECO*, 227). Indeed, argues Veblen, "only when the gifted Jew escapes from the cultural environment created and fed by the particular genius of his own people, only when he falls into the alien lines of gentile inquiry and becomes a naturalised, though hyphenate, citizen in the gentile republic of learning, [does he come] into his own as a creative leader in the world's intellectual enterprise" (*ECO*, 225-26). Aliens in a strange land, Veblen's intellectuals cultivate the supreme knowledge that even the truths they were weaned on are relative and culture-specific. He muses that perhaps this is why the Jews don't proselytize: "the Chosen People have quite characteristically never been addicted to missionary enterprise" (*ECO*, 225). Veblen would have us believe that intellectual Jews' devotion to idle curiosity frees them, as it does him, from the arrogance of impressing upon the heathen an unwelcome final answer. Unlike the dreaded neoclassical economists or the Christians, the Jews do not coerce.

For all of Veblen's protestations about "eliminat[ing] all bias of personality from the technique or the results of science or scholarship" (*HLA*, 7), he slants his discussions of idle curiosity and heroic intellectuals to explain why he felt his own ideas were marginalized: "Any inquiry which does not . . . corroborat[e] . . . opinions in vogue . . . is condemned" (*HLA*, 181). One of the ironies of Veblen's characterization of himself as an outsider and of his social theory as an alien product is that "the outrage expressed by most of the critics" helped to launch his publishing career.[42]

Dreiser more deliberately promoted the view of himself as an outsider in order to help market his wares. As H. L. Mencken astutely told his friend in 1920, commenting on the beneficial effects of the moral outcry against Dreiser's work, "I am thoroughly convinced that the Comstock explosion in the long run was very profitable to you."[43] Dreiser's persona as exemplary artist closely resembles the Veblenian

---

42. Dorfman, "New Light," 16.
43. H. L. Mencken to Dreiser, December 1, 1920, in *Dreiser-Mencken Letters: The Correspondence of Theodore Dreiser and H. L. Mencken*, 2:411. In "Thinking Sex: Notes for a Radical Theory of the Politics of Sexuality," Gayle Rubin describes Anthony Comstock, founder of the New York Society for the Suppression of Vice, as an "ancestral

scientist-scholar marked by skepticism and motivated by idle curiosity; both acquire a legendary force while being self-protective. It is as much a point of professional honor for Dreiser as for Veblen *not* to fit into any mold.[44]

Echoing a key term in the Veblen legend (and anticipating Said's characterization of the intellectual), H. L. Mencken once described Dreiser as "an alien to his place and time." Dreiser certainly liked to advertise the intransigence of his work. He resisted the label most often assigned him: literary realist. As he puts it in a 1930 essay, "I accept the insult [of being a realist] but with reservations. For I fear I do not run true to type—do not march with any clan." Dreiser tended also to resist political categorizations. After his trip to Harlan County, Kentucky, to support striking miners, an event that marked, in Richard Lingeman's words, "the high point of Dreiser's romance with communism," Dreiser said, "I have been told over and over that I am much too much of an individualist and that I do not subscribe to the exact formulas [of] the Communist program, and therefore I would never be admissible."[45] Again, he opens *Dreiser Looks at Russia* by declaring himself "an incorrigible individualist—therefore opposed to Communism" (9). Dreiser articulates the battle cry of the professional outsider as eternally inadmissible.

Dreiser's central strategy for glorifying marginality is to define his writing as working against the grain and to call attention to resistance to his work. In doing so, he cashes in on some of the allegations against

---

anti-porn activist" (4). The 1873 Comstock Act made it illegal to buy, sell, or produce materials considered obscene.

44. In his introduction to Dreiser's *American Diaries 1902-1926,* Thomas P. Riggio writes that Dreiser "created a cultural myth, larger than his own experience, which served as a sustaining metaphor of artistic integrity" (6). Similarly in "A Portrait of the Artist as a Young Actress: The Rewards of Representation in *Sister Carrie,*" Barbara Hochman observes that "Dreiser's accounts of inspired creativity, calculated in part to attract the attention of readers . . . , also serve to dramatize the writer's alleged independence" (44). On Dreiser's independent stance as a marketing device, see Kaplan, *Social Construction;* Wilson, *The Labor of Words;* and Borus, *Writing Realism.*

In *Forgotten Frontiers,* Dudley made much of Dreiser's being an outsider because of his German ancestry, as had Mencken before her. Indeed, one of the bases for the close bond that Mencken and Dreiser developed was their shared German background, as a perusal of their letters—particularly during the war years—reveals. On Dreiser's sense of ethnic marginality see Thomas P. Riggio, "Theodore Dreiser: Hidden Ethnic," and Arthur D. Casciato, "How German Is *Jennie Gerhardt?*"

45. H. L. Mencken, "The Creed of a Novelist," 761; Dreiser quoted in endnote to "True Art," 156; Richard Lingeman, *Theodore Dreiser: An American Journey, 1908-1945,* 362.

him, declaring that his supposed lapses of taste, judgment, and style are, in fact, virtues. He concedes what his angriest critics charged—that his works are morally suspect and incendiary. Much as Veblen insinuated regarding David Hume's reputation in England, Dreiser charges it is not he but America that is hostile, America that has always marginalized, if not ridiculed, its freethinkers.

Like Veblen's aligning himself with a personal gallery of recalcitrant heroes, Dreiser declares that the country's finest authors have always been misunderstood: "Poe, Hawthorne, Whitman and Thoreau, each in turn was the butt and jibe of unintelligent Americans," and "[t]he few genuine thinkers that America has thus far produced are taboo" (*HRDD*, 276, 57). The "mentally unemployed" (*DLR*, 20) throughout the United States have little use for genius. Dreiser's interpretation of America's hostility to creativity lets him justify himself as a critic and blame his war with moralistic critics and narrow-minded readers on his antagonists. As Dreiser puts it, "Personally my quarrel is with America's quarrel with original thought" (*HRDD*, 273). Dreiser suggests that he is not behaving belligerently; he is only defending himself against predictable assaults.

Dreiser offers various explanations for this fierce anti-intellectualism, from leveling the familiar charge that America expends its energy in materialistic pursuits to blaming Mencken's favorite culprit, an excess of puritan morals (*HRDD*, 258, 271).[46] But the deeper problem that Dreiser locates involves the American passion for conformity. In *A Hoosier Holiday,* he generalizes that "everywhere west of Pennsylvania. . . . The idea of doing anything original is severely frowned upon. Whatever else you may be in America or elsewhere, apparently you must not be different" (136). He similarly describes his brief foray into the higher learning at Indiana University: "All were here, presumably, to think original thoughts, or at least to make the attempt, yet if anyone had really dared . . . !" (*D*, 409). Dreiser approvingly quotes the judgment of one commentator that colleges produce "little more than types, machines, made in the image and likeness of their college. They do not think; they cannot, because they are held hard and fast by the iron band of convention" (*HRDD*, 260). Because of this hysterical fear of independence—not any untoward actions of Dreiser's—

---

46. On the Dreiser-Mencken alliance, see Thomas P. Riggio's commentaries in *Dreiser-Mencken Letters.* Particularly significant for my argument is how Mencken "identified Dreiser's work as a form of cultural aggression" (1:182).

nonconformists "are truly terrible to the mass—pariahs, failures, shams, disgraces" (*HH,* 285).

As confirmed recently in the popular reaction against negative political advertising, people quickly grow suspicious from hearing antagonists define each other. Dreiser makes his opponents look terrible by grimly portraying the American psyche as narrow and fearful, yet must still account for how any individual—even a genius such as himself—could escape the clutches of conformity and coercion. Dreiser's cultural criticism depends as much as Veblen's on the proposition of his intrinsic difference. Both consider recalcitrance a precondition for brilliance.

To establish himself as one of those "occasional individual[s] who may rise in spite of these untoward conditions . . . to understand . . . life as it really is," Dreiser combs his memory for early signs of his uniqueness (*HRDD,* 266). The emphasis on his singularity that runs throughout Dreiser's autobiographies typically builds up his public persona as a cultural critic, while the competing strain of wanting to fit in generally emerges in his descriptions of private and personal relationships.

The wave of feminism that swept the United States in the 1970s is generally credited with promulgating the now-axiomatic idea that the personal is political. Dreiser's books may seem an unlikely prologue to modern feminist thought, but they anticipate the assumption that personal experience has political repercussions (and vice versa) and that the private and public realms always interpenetrate one another. In fact, Dreiser's cultural criticism depends on this assumption. His public persona as iconoclast and outsider derives from, and no doubt in part assuages, his haunting sense of being a personal misfit. Through the persona of politically engaged cultural critic, Dreiser made a professional virtue of what felt like personal defects.

The end of *A Book about Myself,* the longest of his tales of apprenticeship, illustrates this Dreiser paradigm. After a promising start as a newspaper reporter in Chicago, St. Louis, and Pittsburgh, a young Dreiser finds New York City impenetrable. But *A Book about Myself,* written twenty-six years after the fact (and published another two years later), transforms what seemed at the time like resounding defeat into a dramatic turning point in his career. Dreiser remembers being "[h]aunted by the thought that I was a misfit" among the successful young men who reported on New York City news. He knew he was a different breed from these bustling youths he describes as professional

"Yea-sayers." At the end of *A Book about Myself* he muses, "Perhaps . . . life as I saw it, the darker phases, was never to be written about" (489, 488, 491). This suggestion of an unvoiced yet more accurate story waiting to be told provides a dramatic setup for the Dreiser legend of a genius beleaguered but triumphant.[47]

As Dreiser mythologizes his career, it begins not with the publication of *Sister Carrie* in 1900 but with the alleged "suppression" of that book by a wary and puritanical publisher. Dorothy Dudley, an important disseminator of Dreiser's version of the now-legendary publication history of his first novel, describes "The villain" as "Propriety"—draped in the gown of Mrs. Frank Doubleday. In her (Dreiserian) version of the story, Dudley even invents outraged dialogue between Mrs. and Mr. Doubleday concerning the immorality of *Sister Carrie*. Dudley's assertion that "the myth has legs to stand on" has been substantially discredited, but Donald Pizer nicely balances the facts against the significance of the legend to literary history: "whatever its insubstantiality in fact, the legend of the suppression of *Sister Carrie* has an independent reality and significance in American cultural history. . . . The facts of the publication of *Sister Carrie* are therefore one kind of truth, the myth of its suppression is another."[48] In fact, Dreiser's surefooted handling of *Sister Carrie,* both before and after its publication, show him at least as eager to establish his voice as a critical force to be reckoned with as to write realistic fiction.

47. On Dreiser's newspaper career, see Shelley Fisher Fishkin, *From Fact to Fiction: Journalism and Imaginative Writing in America*, chap. 4; W. A. Swanberg, *Dreiser;* Richard Lingeman, *Theodore Dreiser: At the Gates of the City, 1871-1907;* Moers, *Two Dreisers;* Elias, *Theodore Dreiser;* T. D. Nostwich's "Historical Commentary" to *Theodore Dreiser Journalism*, vol. 1; T. D. Nostwich's introduction to *Theodore Dreiser's "Heard in the Corridors" Articles and Related Writings;* Yoshinobu Hakutani's introduction to *Selected Magazine Articles of Theodore Dreiser;* Nancy Warner Barrineau's introduction to *Theodore Dreiser's Ev'ry Month.*

48. Dudley, *Forgotten Frontiers,* 171, 173, 181; Donald Pizer in beginning of Legend section of Norton Critical Edition of *Sister Carrie,* 456. Pizer reviews the publication history in his introduction to *New Essays on* Sister Carrie, concluding that "despite his difficulties with Doubleday, Page, Dreiser now considered himself principally a novelist" (13). Dreiser as likely considered himself an important author because of, not despite, conflicts with his publisher. As Christopher P. Wilson observes, "The issue rapidly became not [*Carrie*'s] immorality but whether Dreiser would submit to the firm's management of his literary career" ("*Sister Carrie* Again," 289).

In "The Publication of *Sister Carrie:* Old and New Fictions," Stephen C. Brennan accuses the editors of the Pennsylvania edition of creating yet another "popular myth." On the *Carrie* legend, see also Swanberg, *Dreiser,* and Jack Salzman, "The Publication of *Sister Carrie:* Fact and Fiction."

Dreiser's correspondence regarding the publication of *Sister Carrie* belies his self-description as a mere fledgling novelist, "green as grass." In a 1900 letter to his friend Arthur Henry, Dreiser indicates his intent to force Doubleday to publish *Carrie* and says that he already foresees his fame: "If when better known and successful I should choose to make known this correspondence [from Doubleday and others regarding *Carrie*], the house of Doubleday would not shine so very brightly." Dreiser plays the genius card yet more dramatically in a 1900 letter to Walter Page: "A great book will destroy conditions"—and so Dreiser demands that *his* great book be published.[49]

The Dreiser legend can be read throughout debates over his works that had the "honor," as Mencken puts it, to be suppressed. Dreiser repeatedly capitalized on controversies about both his writing and his life to advance his persona as rebellious cultural critic. A curiously unguarded passage from *A Hoosier Holiday* suggests how thoroughly what Leslie Fiedler describes as Dreiser's "instinct for self-dramatization" could lead him to conflate life with art.[50] Here Dreiser presents himself as, simultaneously, the book's subject, publicity agent, and celebrated author: "In A. D. 1915, Theodore Dreiser, accompanied by one Franklin Booth, an artist, visited the site of this bridge . . . preserved now in that famous volume, entitled 'A Hoosier Holiday,' by Theodore Dreiser" (*HH*, 216).

But Dreiser's career as professional outsider extends well beyond the pages of specific books or local controversies to invest him with the aura of exemplary and mythical artist. In 1920, Mencken wrote Dreiser that "you are not only permanently secure; you have become a sort of national legend. The younger generation is almost unanimously on your side." Mencken's glowing essay for *Smart Set* in 1914 identifies Dreiser's intransigence with the future of American letters: "so long as Dreiser keeps out of jail there will be hope" for United States literature. Mencken also suggests that we read Dreiser himself like a book: "the whole story of the adventures of his books would make a novel in Dreiser's best manner—a novel without the slightest hint

---

49. Dreiser to H. L. Mencken, May 13, 1916, in Norton Critical Edition of *Sister Carrie*, 428 (see also Dreiser to Fremont Older, November 27, 1923, ibid., 459); Dreiser to Arthur Henry, July [23], 1900, ibid., 441; Dreiser to Walter H. Page, August 6, 1900, ibid., 449.

50. H. L. Mencken, "Adventures among the New Novels," 750; Leslie Fiedler, *Love and Death in the American Novel*, 252.

of a moral. His own career as an artist has been full of the blind and unmeaning fortuitousness that he expounds." And, as Mencken says in a backhanded complement in his eulogy for his friend, Dreiser "was more interesting than any of his books."[51]

If Dreiser's quarrel is with America's restraint of free thought, his resolution is that the artist be allowed to do whatever he wishes and depict the world as he finds it. He roundly condemns all "interference" with literary and scientific freedom as "espionage" and pleads, "[t]he artist, if left to himself, may be safely trusted to observe, synchronize and articulate human knowledge in the most comprehensive form" (*HRDD*, 275, 276). Dreiser makes it sound as if "the artist" will master and synthesize all the natural and social sciences. Dreiser's plea that the artist be left alone is a personal dream that he elevates to a principle of cosmic proportions. To be an individual "in the true sense of that word," he argues, one must think independently (*HRDD*, 261). The United States boasts inventors but practically lacks "the supreme freedom of the mind" (*HRDD*, 271). To make the nation truly great will take "thought—intelligent, artistic, accurate vision" (*HRDD*, 58).

In their glamorization of intellectual work as oppositional, some fault lines in Dreiser and Veblen's critical styles emerge. Without resistance from (in Veblen's case) orthodox economists and university bureaucrats or (in Dreiser's) timid puritans and moralists, they would lose the status of intellectual bad boys necessary to their authority. According to their own logic, Veblen and Dreiser must *not* be understood or appreciated by the mainstream culture; that rejection confirms their acuity as critics. But as many contemporary cultural critics are rediscovering, the resistant persona is implicated in the culture under attack. While critical intellectuals today are unlikely to resurrect Veblen and Dreiser's methods for contesting authority, the problems that their rhetorical strategies were intended to address still have not been resolved.

Dreiser's visionary artist and Veblen's skeptical scientist both sabotage the status quo. By redefining truth, attacking institutionalized Christianity, and romanticizing the intellectual's marginality, they made

51. Mencken to Dreiser, October 11, 1920, in *Dreiser-Mencken Letters*, 2:401; Mencken, "Adventures," 749, 751; H. L. Mencken, "A Eulogy for Dreiser," 805. In his 1930 Nobel address, "The American Fear of Literature," Sinclair Lewis seconded Mencken: "without [Dreiser's] pioneering I doubt if any of us could, unless he liked to be sent to jail, seek to express life and beauty and terror" (8).

cultural criticism look smart, sexy, and appealing. Their rhetoric of confrontation decisively influenced their reception. In 1935, Max Lerner identified the authority of Veblen's insubordination: "He was, like Marx and Nietzsche, a symbol by which men measured their rejection of the values of the established order." And in 1932, Dorothy Dudley made a similar claim for Dreiser: "All of those who could not dare but wished they might, found solace in Dreiser. And those born to dare found precedent in him. He became a source of thought and of work."[52]

They sought to distinguish themselves from what Veblen referred to as the "national mob-mind" and Dreiser called that "dread . . . conventional point of view" (*AO,* 39; *HH,* 113). Throughout his works, Dreiser makes conventionality look as unappealing as possible. Consider, for instance, one of his most dismissive fictional characterizations, Frank Cowperwood's first wife, Lillian:

> The conventional mind is at best a petty piece of machinery. It is oyster-like in its functioning, or, perhaps better, clam-like. It has its little siphon of thought-processes forced up or down into the mighty ocean of fact and circumstance; but it uses so little, pumps so faintly, that the immediate contiguity of the vast mass is not disturbed. Nothing of the subtlety of life is perceived. (*F,* 216–17)

To follow conventional wisdom, in other words, is to prove oneself no more evolved than a clam. One of Veblen's most outlandish images uses a bivalve to make essentially the same point. In *On the Nature of Peace and the Terms of Its Perpetuation* (1917), Veblen's brilliant analysis of how patriotism is drummed up during times of war, he ridicules jingoistic nationalism:

> The analogy of the clam . . . may at least serve to suggest what may be the share played by habituation in the matter of national attachment. The young clam, after having passed the free-swimming phase of his life, as well as the period of attachment to the person of a carp or similar fish, drops to the bottom and attaches himself loosely in the place and station in life to which he has been led; and he loyally sticks to his particular patch of oose [*sic*] and sand through good fortune and evil. It is, under Providence, something

---

52. Max Lerner, "What Is Usable in Veblen?" 130; Dudley, *Forgotten Frontiers,* 367. Veblen so influenced a generation that one of his books—not *Leisure Class* but *Business Enterprise*—was included as one of twelve entries in *Books That Changed Our Minds* (1940), ed. Malcolm Cowley and Bernard Smith.

of a fortuitous matter where the given clam shall find a resting place for the sole of his foot, but it is also, after all, "his own, his native land" etc. (*NOP,* 134)

The bizarre personifications—"the person of a carp or similar fish," the punning "sole of [the clam's] foot"—only make Veblen's patriotic clam more hilariously memorable.

Dreiser and Veblen led separate battles in the same war on conventional pieties that raged in the United States during the early twentieth century. Dreiser's persona as cultural critic informs his novels through his consistent identification with those characters he sees as marginal. He generally correlates marginality with economic status; his marginal characters tend to occupy either end of the economic spectrum. Dreiser's Jennie Gerhardt and Frank Cowperwood—the poor, second-generation American who works as a maid, and the financial wizard who makes millions—come the closest to translating into fiction Dreiser's own persona as the outsider who challenges convention. The more familiar characters who exist in, or aspire to, the middle class—such as Carrie Meeber, George Hurstwood, Clyde Griffiths, and Roberta Alden—often experience the pain of social exclusion, but they long to become (or remain) conventional insiders in a way that Jennie Gerhardt and Frank Cowperwood never do. Dreiser's sympathy for characters who live as if conventions don't apply to them can take precedence over his well-known sense of economic determinism. For Dreiser, extreme wealth or poverty can produce the marginalization that he admires.

For Veblen, it is inconceivable that one could be rich and marginal, or poor and nonmarginal. More consistently the economic determinist than Dreiser, Veblen considers economic class the defining factor in personality structure. His crucial distinction between business and industry leads him unequivocally to condemn anyone with status claims, whether based on money, pedigree, or occupation. In the next chapter, I will take up the business/industry distinction, which operates in Dreiser's *Trilogy of Desire* and grounds Veblen's social theory. Yet because Dreiser perceives the financial tycoon as a rebel against convention, even as something of a cultural critic, he glorifies him in a way that would be anathema to Veblen.

2

Business as (Un)usual

The Immaterial Economy in *The Trilogy of Desire*

The long-standing fascination that literary and cultural critics, particularly those in the Marxist tradition, have had with the market has recently been reaffirmed by the popularity of new historicism. Howard Horwitz explains the current charm that economics holds for literary critics as a matter of discovering the transferability of ideas basic to one field into another: "All acts of representation (all rhetorical figures) are acts of exchange—one sign standing for another." Yet, apart from treatments by Horwitz and Walter Benn Michaels, the wave of interest in the literature/economic nexus has barely touched Dreiser's *Trilogy of Desire*, arguably the most sustained fictional representation of economics written by a United States author.[1] This chapter implicates Veblen's economic theory, as developed especially in *The Theory of Business Enterprise* (1904), with Dreiser's *Trilogy*, spotlighting the very different ways they see cultural criticism acting upon (or in) the market.

What I will pursue here confirms the transferability of ideas between economics and literature but focuses on unpacking a series

---

1. Horwitz, *By the Law of Nature*, 15; Michaels, *Gold Standard*. Both emphasize aspects of what I describe, following Veblen, as the immaterial basis of business. Michaels notes that "love of abstraction is central to the financier's career" and sees Cowperwood exploiting a "quirk of identity" in money "that makes it possible to transcend . . . 'actual fact' " (ibid., 64, 67). Horwitz traces the immateriality to a specific practice that Cowperwood favors, hypothecation (whereby he pledges as security papers that he does not in fact own) and to what he calls the hypothecated self (*By the Law of Nature*, especially 196–201).

of theoretical confrontations between Veblen and Dreiser. Veblen's critical distinction between "business" and "industry," which Joseph Dorfman describes as the "fundamental antithesis" that structures all of his thinking, illustrates again his distinctive use of a social scientific concept as a weapon for cultural criticism.[2] Veblen's various and extended critiques of capitalism all derive from his "scientific" separation of business and industry, which he uses to argue for the scandalous immateriality of business and to discredit the work of financiers. The immateriality that Veblen attributes to business incorporates, as we will see, the fictitious constructs of economic orthodoxy that defend what they purport to examine, the superstitious deference to capitalism embodied in platitudes such as "what's good for business is good for America," and such familiar illustrations of the insubstantial nature of capitalism as the volatility of stock value. In *The Financier* (1912), *The Titan* (1914) and *The Stoic* (published posthumously in 1947), Dreiser also seizes upon the idea of the immaterial basis of business, but he transforms its associations into positive ones, as seen in his insistence on the inscrutability, ephemerality, aesthetic possibility, and what he often refers to as the "subtlety" of speculative finance. But Dreiser, far from criticizing modern business, presents it as a metaphor for—and in some respects even a model of—the work of the cultural critic.

That Dreiser uses a foundational Veblenian concept, even while overturning it, becomes particularly clear in his characterization of Frank Cowperwood. Critics have traditionally been embarrassed by how Dreiser combines a realistic transcription of economic process with a romanticized tribute to a personality. But that apparent contradiction dissolves once we grant the accuracy of H. L. Mencken's claim that Cowperwood "fits into no *a priori* theory of conduct." The bristling nonconformity that Mencken saw in Cowperwood is, as we have seen, a key element of Dreiser's own persona as a cultural critic. Far more than he does even with Eugene Witla of *The "Genius,"* Dreiser projects a fantasy image of himself onto Cowperwood.[3] Dreiser sees

2. Dorfman, *Thorstein Veblen*, 239.
3. Mencken, "Adventures," 753. In "Cowperwood's Will to Power: Dreiser's *Trilogy of Desire* in the Light of Nietzsche," Joseph C. Schöpp similarly identifies Cowperwood with Dreiser but argues that the identification proceeds from the novelist's use of Nietzschean thought. For a contrasting reading, see Philip L. Gerber's contention in

a logical connection between his character's having the courage to flout conventional and institutional strictures and his mastering of the immaterial world of business. Dreiser depicts Cowperwood as a figure whose actions, thoughts, desires, and very personality critically defy his culture's pieties, whereas Veblen would see the likes of Cowperwood as a fit subject for criticism to dismantle.

My goals for this chapter are threefold. I want, first, to lay out the central tenets of Veblen's economic theory, which explain a great deal about capitalism's durability and seductive power over the American imagination. Veblen's analysis of business enterprise provides a useful alternative to Marxist theory that is, nonetheless, in many respects compatible with it.[4] Second, I will show Dreiser's mastery of economic processes by correlating his explanation in the *Trilogy* with Veblen's analysis of capitalism. I want thereby to lay to rest the critical tendency to underrate Dreiser's attempts at socioeconomic theorizing.[5] My third goal is to use the disparity in Dreiser's and Veblen's judgments on the immateriality of business to examine further their respective positions as cultural critics. Dreiser's and Veblen's conclusions about the financier's personal and social value are particularly interesting because they move from the same premises to radically different conclusions. Veblen's and Dreiser's disagreement involves matters of no less consequence than the role business plays in American society, be it destructive or creative, revolutionary or reactionary. On this same ground many political and social policy battles continue to be fought in the United States.

Theodore Dreiser that Dreiser created the financier "[b]y inverting his own personality," making Cowperwood "as unlike himself as possible" (93).

4. Veblen's relationship to Marx, a particularly difficult question, deserves protracted attention that is not possible here. Veblen's three essays on socialism provide a good starting point (*POS*, 387–456). On the relationship between Marx and Veblen, see Diggins, *Bard;* Tilman, *Thorstein Veblen and His Critics;* Robert Griffin, *Thorstein Veblen: Seer of American Socialism;* and Leonard Dente, *Veblen's Theory of Social Change.*

5. For an early and definitive statement of Dreiser's mushy philosophizing, arguing that the fiction transcends Dreiser's often inept pronouncements, see Eliseo Vivas, "Dreiser, an Inconsistent Mechanist." In *The Art of Frank Norris, Storyteller,* Barbara Hochman notes a similar phenomenon in Frank Norris, whose novels she finds more capacious and interesting than the overtly "naturalistic" pronouncements of the narrators. On Dreiser's philosophizing, also see Martin, *American Literature,* 218–19; Warren, *Homage,* 90; Zanine, *Mechanism and Mysticism,* especially 174; and Riggio's introduction to Dreiser's *American Diaries,* 20.

Business versus Industry

Two quotations encapsulate the difference in Veblen's and Dreiser's judgments on business. In *The Vested Interests and the Common Man* (1919), Veblen describes "business as usual": "Already the vested interests . . . are busily arranging for a return to business as usual; which means working at cross-purposes as usual, waste of work and materials as usual, restriction of output as usual, unemployment as usual, labor quarrels as usual, competitive selling as usual, mendacious advertising as usual, waste of superfluities as usual by the kept classes, and privation as usual for the common man" (140-41). This passage concisely illustrates Veblen's view that to conduct business means to profit by plundering the community's resources, to benefit from creating friction and inefficiency. It is an incisive, but hardly a dispassionate, characterization: Veblen grants financial tycoons excellence in no less than lying, cheating, stealing, and of course in his pet vice, wasting. Dreiser's contrasting position in "The American Financier" (1920), no less passionate, affirms "the most useful of all living phenomena" (*HRDD*, 74) to be the tycoon:

> At best, all we have is the individual, not always financial, by any means, or artistic, but one who has dreamed out something: music, a picture, poetry, a machine, a railroad, an empire—anything, in short, that man as race or nation can use or rejoice in. If to have a Woolworth Building, a transcontinental railroad, a Panama Canal, a flying machine, to say nothing of literature and art, means that we must endure a man who is dull, greedy, vain, ridiculous . . . [t]hen let us have him. (*HRDD*, 90)

Although Dreiser grants that tycoons may exhibit the dullness and greed Veblen attributes to them, he emphasizes instead their instrumentality, artistry, and individuality. Dreiser is neither indifferent toward nor naive about the predatory and obstructive tactics of capitalism of such concern to Veblen—when he writes elsewhere of the "idle, greedy, or predatory rich," Dreiser's values coincide precisely with Veblen's (*HRDD*, 43-44)—but when it comes to evaluating the rare individuals he sees as succeeding in business by defying their culture's conventions, Dreiser is after other game.

Dreiser's fusing of artistic creativity with business success, which informs the *Trilogy of Desire* as well as "The American Financier," suggests a great deal about his personal investment in Cowperwood's intransigence. Dreiser extends the idea also to emphasize the financier's productivity. According to Dreiser, even the tycoon's greed and vanity ultimately benefit the community by leading to the creation of things people can use: buildings, railroads, canals, airplanes. Not simply a benefactor of humanity, the titanic businessman is also, according to Dreiser, one of the elect. Most of us, he asserts in an important passage in *The Financier,* "do not know what it means to be a controller of wealth, to have that which releases the sources of social action—its medium of exchange" (182). In contrast with average people who "want money, but not for money's sake. They want it for what it will buy in the way of simple comforts," the financier desires money "for what it will control—for what it will represent in the way of dignity, force, power" (*F,* 182). Business carried on at a certain level is clearly, to Dreiser, *un*usual.

This difference in judgment on "business as usual" derives from Dreiser's and Veblen's contrary evaluations of the competing interests of business and industry.[6] According to Veblen, the "economic situation is a composite of industry and business" (*AO,* 256), and there is no mistaking which component he prefers. Veblen defines business as efforts undertaken for "pecuniary gain" which do not, "either proximately or remotely," lead to the production of goods, which is the task of industry (*TBE,* 28; *POS,* 293). Businessmen pursue an "alteration of the distribution of wealth" rather than "productive or industrial activity"; they manipulate money and other emblems of value rather than making anything useful (*POS,* 296). He claims that prosperity, once measured in terms of "industrial sufficiency" or the number of goods produced, has degenerated into the businessman's "quest of profits" (*TBE,* 178). Industry champions "serviceability" but business only the "vendibility" of goods (*TBE,* 51). Veblen sharply contrasts the utilitarian, instrumental productions of industry with the self-seeking, ceremonial manipulations of business.

Veblen even doubts that the nineteenth-century hero, the so-called Captain of Industry still exists; he sees around him mere "Captain[s] of Solvency" and "Lieutenant[s] of Finance" (*AO,* 114; *EPS,* 81). As examples of this new breed of warriors, he cites real estate agents, bankers,

---

6. A concise overview of the business/industry distinction can be found in Veblen's "Industrial and Pecuniary Employments."

brokers, and (to irritate even more people) attorneys (*POS,* 293). But the people who best exemplify what Veblen defines as businessmen are the "financiering strategist[s]," "pecuniary experts" and financial "undertakers" (*TBE,* 22, 29; *POS,* 288) such as Cowperwood, who makes his first million speculating in the stock market, later moving into public services like natural gas and the street railway and subway systems. Cowperwood never concerns himself with production but always with what Veblen calls "the higgling of the market" (*VI,* 89; *POS,* 294).

In *The Financier,* Dreiser suggests the difference between business and industry in a description of Edward Malia Butler, the father of Cowperwood's girlfriend and an ally of the financier who later becomes his foe. Butler once performed an industrially useful function—he collected garbage—but by the time of the action of *The Financier,* he has contracted the work out, "mak[ing] between four and five thousand a year, where before he made two thousand" (65). A remark by Cowperwood's father early in *The Financier* that "if we had a bundle of those [stocks in the British East India Company] we wouldn't need to work very hard" also indicates the distinction between industry and business (11). These references to the pecuniary advantages of *not* performing socially useful work illustrate Veblen's claim that business is far more lucrative than industry. Indeed, while industry entails making things, business means making money, and so Veblen concludes "[t]he highest achievement in business is the nearest approach to getting something for nothing" (*POS,* 303; *VI,* 92). The financier's only activity in *The Trilogy of Desire* that Veblen would consider industry occurs during Cowperwood's incarceration, when he canes chairs. (Appropriately enough, Veblen affirms that the "interest in work differentiates the workman from the criminal on the one hand, and from the captain of industry on the other" [*TLC,* 242].) At all other times Frank Cowperwood is a "pecuniary expert" par excellence.

Veblen's invidious distinction between industry and business discloses a fundamental ethical position that grounds his entire social theory and cultural criticism. He aligns industry with the "instinct of workmanship," which causes people to want to make things, and technology. According to Veblen, technology forms a "living structure" that continually evolves, reflecting the "joint stock of knowledge derived from past experience" and "held in common by the civilized peoples"

(*AO*, 65; *EPS*, 56, 82).⁷ The communal quality is of prime importance: technology unites people through time vertically (the present generation benefiting from the technological wisdom of all prior generations) and horizontally (all citizens at a given historical moment inheriting the same technological knowledge). The "substantial core" of any civilization, a community's technological knowledge binds it together (*VI*, 57; *POS*, 325). But in contrast with the collective bent of industry, business grows out of the institution of private property and pursues only an "advantageous discrepancy between the price paid and the price obtained" (*TBE*, 152). The crucial word is *discrepancy:* whereas the goal of industry (maximizing production) tends to benefit everyone, the goal of business (differential advantage) causes conflict and competition. Notwithstanding Veblen's veneration of the individualistic bent of idle curiosity, his social values fall squarely on the side of communalism.

The real scandal, according to Veblen, is that business and industry have become so muddled. He excoriates that "[i]ndustry is carried on for the sake of business, and not conversely" (*TBE*, 26). Indeed, the "distinguishing mark of any business era" is the infiltration of industry by "pecuniary principles" (*IOW*, 216). He sees modern business as "industrially parasitic" (*TBE*, 64). That judgment informs one of Veblen's memorable phrases, the "usufruct of the state of the industrial arts" (*VI*, 69).

The vantage point of Veblen's criticism becomes yet more sharply defined when he analyzes how businessmen produce differential advantages so as to make money. He claims that profits result from obstructing industry, curtailing production, and wasting resources. Although the businessman is personally "indifferen[t]" to whether he "help[s] or hinder[s] the system," his end being simply "pecuniary gain," often the capitalist finds the greatest profit lies in "disturbance of the industrial system" (*TBE*, 29, 28). Indeed, since the businessman seeks differential advantages, "obstructive tactics" against competitors, such as the "'freezing-out' of rival concerns," can be highly lucrative

---

7. See also Karl Marx: "Every productive force is an acquired force, the product of former activity. The productive forces are therefore the result of practical human energy; but this energy is itself conditioned by the circumstances in which men find themselves, by the productive forces already acquired, by the social form which exists before they do, which they do not create, which is the product of the preceding generation" (1846 letter, excerpted in *Marx-Engels Reader*, 137).

(*POS*, 355). Simply put, the wise business decision frequently pivots on "disserviceability" (*POS*, 356). Vivid examples of the disserviceability of business to the general population can be drawn from the career of Charles T. Yerkes, Dreiser's primary historical model for Cowperwood. Veblen was no doubt familiar with Yerkes: during the late 1890s, when the robber baron was fighting for his Chicago traction franchise, Veblen was teaching at the University of Chicago. Yerkes's Chicago trolley cars were frequently overcrowded, unventilated, and dangerous. Forty-six people were killed and another 336 injured in one year due to improperly installed overhead wires. Yet when his own stockholders pleaded with him to make repairs, Yerkes curtly replied, "It is the people who hang to the straps who pay you your big dividends."[8] As Captains of Solvency such as Yerkes can profit from disserviceability, so can they benefit from large-scale financial panics that "leav[e] the business men collectively poorer, in terms of money value; but the property which they hold between them may not be appreciably smaller in point of physical magnitude or mechanical efficiency" (*TBE*, 191). As Christopher Shannon succinctly describes Veblen's point, "that financial gain can result from productive loss is not an 'irony' of modern business; it is its motor principle."[9] The crucial fact for Veblen is that the industrial basis ("physical magnitude or mechanical efficiency") remains relatively stable during financial crises; a "pecuniary, not a material, shrinkage" occurs (*TBE*, 191).

Although Cowperwood embodies business, not industry, he exploits the distinction between the two. When business looks bad, Cowperwood reminds his creditors of the solid industrial basis beneath him. For instance, when the Chicago fire sets off a financial panic that gives Edward Malia Butler an excuse to call in loans made to his daughter's seducer, Cowperwood finds himself needing cash. Trying to extract more money from the city coffers, Cowperwood explains to the cowardly Philadelphia city treasurer, George Stener, that the decline in selling price has not affected the long-term value behind their stocks, for "the railways are [still] there behind them" (*F*, 204). Even when forced by his creditors to close up shop, Cowperwood still insists that the problem reflects business rather than industrial conditions: "There is nothing the matter with the properties behind them [his securities]" (*F*, 218).

---

8. Sidney I. Roberts, "Portrait of a Robber Baron: Charles T. Yerkes," 352.
9. Shannon, *Conspicuous Criticism*, 14.

Understanding that his profit or loss has little connection with the actual industrial value of his holdings, when Cowperwood can cash in on industry, he happily does so.

Take, for instance, the American Match episode, one of the most dramatic sequences in the *Trilogy*. Chicago's powerful "quadrumvirate" (*T,* 383)—Hand, Schryhart, Merrill, and Arneel—pay dearly for their hatred of Cowperwood when their investment in an attempted monopoly of wooden matches goes awry. The quadrumvirate's trouble begins when the political climate turns hostile to business. The 1890s debate over whether gold, silver, or a combination of the two should back the United States dollar dampened financial markets, but when the "Apostle of Free Silver" (*T,* 367), William Jennings Bryan, was nominated for president in 1896, inflationary fears depressed the economy further.[10] Even though American Match is "one of the strongest of market securities," its stock plummets (*T,* 372–73). In a desperate attempt to sustain the market, the two original promoters, Hull and Stackpole, approach Cowperwood for a loan, even though they know the quadrumvirate disapproves of him.

The silver agitation has created a climate favorable to Cowperwood's pecuniary advancement. Veblen explains the general principle: "A convincing *appearance* of decline or disaster will lower the *putative* earning-capacity of the concern below its *real* earning-capacity," thus providing an opportunity for clever investors to buy stocks at bargain prices (*TBE,* 161, emphasis added). Cowperwood seizes the opportunity for immense gain by exploiting the difference between the substantial, "real" value of American Match and its lower "putative" value caused by the depressed market. Cowperwood realizes this gain not only in wealth but also, more important, in what Veblen calls "strategic control"—and, as it turns out, control of Chicago as well as American Match (*TBE,* 161). Stackpole's claim, "There's nothing the matter with this stock. It will right itself in a few months" (*T,* 377), is an accurate

---

10. Cowperwood prefers gold from an early age, as the 1912 *Financier* makes especially clear: "This medium of exchange, *gold,* interested him intensely. He asked his father where it came from, and when told that it was mined, dreamed that he owned a gold-mine and waked to wish that he did. Even what gold was made of—its chemical constituents—interested and held his attention. He marveled that it ever came to be, and how it was finally selected as the medium or standard of exchange" (1912 *F,* 18; the slightly watered-down passage in the 1927 text occurs on p. 11 of the Signet edition).

See Michaels, *Gold Standard,* for an important analysis of the implications of the gold standard debate in American culture.

enough assessment of the industrial basis, yet irrelevant to the conduct of business. Cowperwood declines to loan money to Hull and Stackpole himself, while instructing some of his financial puppets to make the loan. Cowperwood then throws stock on the market, further lowering the price. The quadrumvirate loses money and Hull and Stackpole are ruined.

As Veblen would predict, Cowperwood could not care less about the actual production of matches. Rather than considering the industrial "output of goods as [a] source of gain," Cowperwood's business lies in the "alterations of values involved in disturbances of the balance" (*TBE,* 49). Veblen further explains, "This work lies in the middle, between commercial enterprise proper, on the one hand, and industrial enterprise in the stricter sense, on the other hand. It is directed to the acquisition of gain through taking advantage of those conjunctures of business that arise out of the concatenation of processes in the industrial system" (*TBE,* 49). Cowperwood works in such "interstitial" spaces where, as Veblen remarks, "partial information, as well as misinformation, sagaciously given out" can lead to the desired outcome (*TBE,* 49). Thus Cowperwood declines to take the American Match stock outright, while turning around and buying it indirectly, thereby obscuring his own role in the process.

Cowperwood's machinations over American Match illustrate Howard Horwitz's contention that Dreiser's financier paradoxically consummates his individualism by obscuring his own agency.[11] The episode also illustrates Veblen's claim that spectacular business deals typically proceed by disturbing the economy. In Veblen's words, "pecuniary experts. . . . have an interest in making the disturbances of the system large and frequent," because their profits come from a change in pecuniary values, not the production of more or better goods (*TBE,* 29).

Strategic Mismanagement

An earlier economic crisis depicted in *The Financier* also illustrates Veblen's point about the "disserviceability" of businessmen to the general community (*POS,* 356). Just as the panic following the Chicago fire leads directly to Cowperwood's prosecution, incarceration, and

---

11. Horwitz, *By the Law of Nature,* 197.

(most pointedly) reduction of his assets, the 1873 panic marks his financial resurrection. Seeing "opportunity" where others see "ruin" (*F,* 440), he makes one million dollars by selling short (i.e., selling stock he does not own in anticipation of a decline in price, when he can buy what he needs at a lower rate to cover his sales at a higher price). In doing so, Cowperwood again illustrates Veblen's point that financial killings, as they are so aptly called, derive from a businessman's sabotaging the economic system from which he profits.

What makes the 1873 panic especially interesting is that the failure of another financier, Jay Cooke, brought it about. A banker who helped finance the Civil War by selling government bonds, Cooke moved into railway lines and wanted, as the narrator of *The Financier* puts it, "to connect the Atlantic and the Pacific by steel" (436). Dreiser does not belabor the point, but notes that Cooke "knew little of railroad-building, personally" (*F,* 437) and that his ignorance of the industry he tried to finance contributed to his failure.

Businessmen's ignorance about the industry they "mismanag[e]" forms one of Veblen's key charges against them (*TBE,* 161). Although when it comes to people who work for a living, "imbeciles are useless in proportion to their imbecility," Veblen claims the reverse is true for businessmen (*POS,* 344). Only the "enforced incompetence" of businessmen causes them to keep prices up by limiting production, when more "intelligent control . . . would have made [higher production] commercially profitable" (*EPS,* 78). Were the goal to produce the best goods at the lowest cost, Veblen savagely remarks, "the experienced and capable business men are at the best to be rated as well-intentioned deaf-mute blind men" (*EPS,* 138). His contempt for businesslike irrationality causes Veblen to fashion one of his sharpest oxymorons, to describe the "astute mismanagement" of industry by business (*IG,* 320). (I would offer my own Veblenian phrase: instrumental imbecility.)

Veblen's insistence on the irrationality of modern business should not, however, be dismissed as splenetic satire. His position complements an influential theory of the motivation of capitalists published at the same time: Max Weber's *The Protestant Ethic and the Spirit of Capitalism.* Weber's famous contention, that the "spirit of capitalism" reflects the Calvinist idea of a predestined "calling" for the elect, leads to the equally important idea that modern business is distinguished by its rationalization. That is, Weber contrasts premodern business as "irrational" work, "directed to acquisition by force, above all the acquisition of booty," with the penchant of modern business for rational "calculation of capital in terms of money.... Everything [in modern business] is done in terms of balances."[12] Veblen's claim about the irrationality of modern business does not simply invert Weber's position, for what Weber means by "rationality" has more to do with what Veblen calls "pecuniary accountancy," or the flattening of all evaluation to numbers, than with irrationality (*POS*, 245). Veblen, however, implies that the sort of rational calculation of interest to Weber masks a profound irrationalism: in Veblen's view other yardsticks, such as serviceability, productivity, and usefulness to the entire community, provide more significant ways to measure value than does the pecuniary calculus. In fact, Veblen characterizes the (supposedly rational) reduction of value to numerical values as "businesslike imbecility" (*AO*, 360). Veblen's characterization of modern business as predatory and destructive thus turns on its head Weber's claim about the peaceful, rational nature of modern business.

There are many examples of Cowperwood operating by what Veblen calls pecuniary accountancy, including his attitude toward the Civil War ("It meant self-sacrifice, and he could not see that.... He would rather make money" [*F*, 62]); his response to his children ("He liked it, the idea of self-duplication. It was almost acquisitive" [*F*, 57]); and his rationalization for defecting from his first wife ("He was not doing her any essential injustice, he reasoned—not an economic one—which was the important thing" [*F* 422]). Dreiser also depicts his financier as substantially ignorant about the industry he controls. Knowing "absolutely nothing of the business of gas—its practical manufacture and distribution"—does not stop Cowperwood from making a fortune from it in *The Titan* (46). Admitting "I'm not a practical gas man myself," Cowperwood simply hires Henry De Soto Sippens to run the Chicago company for him, turning into what Veblen refers to as an absentee owner who exerts financial control over a concern that he does not actually manage (*T*, 47; Veblen, *AO*).[13] Illustrating the parasitic relation-

---

12. Max Weber, *The Protestant Ethic and the Spirit of Capitalism*, 20, 18. Weber's book was first published in German (one of many languages that Veblen read) in 1904–1905; Veblen's *Business Enterprise* was published in 1904. On Weber and Veblen, see Diggins, *Bard*.

13. *Sister Carrie*'s George Hurstwood also illustrates the division of management from ownership that Veblen calls absentee ownership. While Hurstwood "looked the part" of a successful "man about town" and radiates "above all, his own sense of his

ship that Veblen rails against, Cowperwood gloats, "We may never need to lay a pipe or build a plant" to turn a profit (*T,* 82).

The Chicago tunnels perfectly exemplify the difference between industry and business: although not "properly built," they nevertheless form a "handsome commercial scheme." The industrially questionable tunnels become even more lucrative when Cowperwood seizes upon technological innovations such as a new traction system, arc lighting, and the telephone to increase their value (*T,* 158, 157, 158). His scheme to gain control of the existing tunnels shows how businessmen translate industrial developments into pecuniary gain.

Cowperwood's orientation in *The Stoic* remains the same. Indeed, Cowperwood now blithely condescends to the railway engineers, elevating the work of financiering over that of industry: "What interests me, gentlemen . . . is that you who appear to understand the engineering business thoroughly should assume the business of financing to be less difficult. For it isn't of course. Just as you have had to study for years, . . . so I, as a financier, have had to do exactly the same thing" (58). Cowperwood's operating assumption is that "a man who knew enough about [railways] to have extracted twenty million dollars out of it must have some definite knowledge" (*S,* 76). Cowperwood lectures the engineers, Greaves and Henshaw, about the "practical work" he does (*S,* 58). As Veblen would gloss this word, " '[p]ractical' in this connection means useful for private gain; it need imply nothing in the way of serviceability to the common good" (*HLA,* 193).

Cowperwood's treatment of the engineers illustrates what Veblen describes as the "paralogisms" (that is, faulty arguments) typical of businessmen (*TBE,* 291). Few contrasts emerge more sharply in Veblen's writing than that between businessmen and engineers. As Warner Berthoff has remarked, Veblen's engineer/financier dichotomy "has the imaginative authority of great fiction."[14] Veblen as quickly defends the "engineer's conscience" for being "responsibl[e] to his own sense of

---

importance," his position is "a kind of stewardship which was imposing but lacked financial control" (Penn *SC,* 43).

According to Veblen, absentee ownership, or the division of ownership from management, began in the mid-eighteenth century but had grown common by the early twentieth (*VI,* 42–43). "The idol of every true American heart," absentee ownership is nevertheless "noxious to the common good" (*AO,* 12; *EPS,* 143). Veblen defines it as the "ownership of an industrially useful article by a person or persons who are not habitually employed in the industrial use of it" (*EPS,* 143).

14. Warner Berthoff, "Culture and Consciousness," 495.

workmanlike performance," as he condemns the "arts of business" such as "bargaining, effrontery, salesmanship, make-believe" (*AO*, 107). In *The Engineers and the Price System* (1921), Veblen flatly declares that Lieutenants of Finance such as Cowperwood "have no technological value, in fact" (127). It is the engineers, he insists, who "make up the indispensable General Staff of the industrial system" and who "alone are competent to manage" it (*EPS*, 82-83). Notwithstanding Max Weber's analysis, a truly rational decision, according to Veblen, would be to let the engineers take over business. Were that to happen, Veblen predicts production would increase—by as much as several hundred percent (*EPS*, 119-20).

Many Dreiser commentators see him manifesting great respect for his fictional engineer, Bob Ames in *Sister Carrie,* a character I will consider in the next chapter. I am not suggesting that the *Trilogy* commits Dreiser unequivocally to Cowperwood's position regarding the superiority of financiers (business) over engineers (industry). The issue worried and confused Dreiser, who throughout his life bewailed his own (and his father's) "inefficiency" and admired any instance of industrial competence. In *A Hoosier Holiday* (1916), for instance, Dreiser indulges in the boasting praise for American technology so common then (and still). Countering the familiar complaint that "from the point of view of patina, ancient memories, and the presence of great and desolate monuments" America seems dull, Dreiser counterpoints, "[but] contrasted with our mechanical equipment Europe is a child. Show me a country abroad in which you can ride by trolley the distance that New York is from Chicago" (*HH*, 61). Dreiser's enthusiasm for technology also led to his friendship with Howard Scott, the engineer who founded the Technocracy group (an organization greatly influenced by Veblen's ideas, especially those expressed in *The Engineers and the Price System*).

Nevertheless, in the *Trilogy,* Dreiser goes out of his way to associate financiers with productive, even creative work. In *The Financier,* Jay Cooke receives kudos for his "constructive work" and "genius"; Dreiser depicts Cooke's railroad ambitions as an attempt to erect "a permanent memorial to his name" (435, 436). In *The Titan,* Aileen contrasts her husband's "constructive persistence" with the sybaritic lifestyle of her beau, Polk Lynde (351). Similarly, in *The Stoic,* Berenice decides against pursuing a romantic involvement with Lord Stane when she realizes he "lacked the blazing force of Cowperwood . . . the fascinating fan fare and uproar that seemed ever to accompany the great in the rush and flare of creation" (202). Like Jay Cooke, Cowperwood dreams to leave London "a modern and comprehensive metropolitan system which would bear the imprint of his genius just as Chicago's downtown loop bore it" (*S*, 214). Dreiser even has the English aristocrat and Cowperwood's business partner in *The Stoic*, Lord Stane, assert the serviceability of the financier: these "real lords of the future . . . would be the greatest factor in society's progression" (69).[15] Cowperwood believes himself to be an "immense creative force" in the "practical realm" (*S*, 164, 189)—and so does Dreiser.[16]

The Epistemology of Finance

Pursuing Dreiser and Veblen's disagreement over whether the Captains of Solvency enact practical, creative changes or incompetent, destructive ones takes us to a second arena of theoretical confrontation, one where the role of cultural criticism emerges sharply. According to Veblen's analysis of the modern economic situation, working in business or in industry conditions different epistemological orientations. Modern industry, he argues, operates by the "machine process" which "gives no insight into . . . good and evil" and disdains "conventionally established rules of precedence." In fact, the machine process rejects "immemorial custom, authenticity, or authoritative enactment" (*TBE*, 311). Making "no use of conventional, sentimental, religious, or magical truths," the machine process cultivates cause-and-effect thinking (*AO*, 280).[17] People who work in fields that feature technological or industrial

---

15. Dreiser anticipates this proposition in one of the most famous passages of *The Financier*: " 'I satisfy myself,' was his motto; and it might well have been emblazoned upon any coat of arms which he could have contrived to set forth his claim to intellectual and social nobility" (121). Dreiser thus fused the idea of "nobility" with Charles T. Yerkes's statement that "Whatever I do, I do not from any sense of duty, but to satisfy myself, and when I have satisfied myself, I know I have done the best I can" (quoted in Roberts, "Portrait," 351). Dreiser's attitude emerged as early as his 1899 interview of Andrew Carnegie (appropriately titled "A Monarch of Metal Workers") for Orison Swett Marden's *Success* magazine.

16. Although there remain conspicuous American megalomaniacs such as Donald Trump who still affirm "the art of the deal," the more common variant of the idea of business as creative force tones down the Dreiserian rhetoric considerably. As the twentieth century draws to a close, one is likely to hear that business "creates" jobs.

17. Veblen explains: "The technology of physics and chemistry is not derived from established law and custom, and it goes on its way with as nearly complete a disregard

changes thus tend to develop immunity to the institutional thinking that molds most people's behavior.

If Veblen were correct, then today's computer gurus should all be freethinking agnostics for whom Nietzsche is bedside reading. However unlikely that scenario, Veblen's formulation remains important, for its terms recall how he stakes out for himself the allegedly dispassionate ground of scientific analysis that, he claims, dodges conventional wisdom (as embodied in imbecile institutions) and defies orthodoxy. In other words, the features that Veblen attributes to the "machine process" correlate with the qualities he associates with cultural critics.

Veblen also analyzes a second epistemological orientation that he claims grows out of capitalism. In doing so, he further sharpens the contrast between his own oppositional values and those he claims typify the business community. He argues that the mindset of businessmen includes the sort of rationalism of such interest to Max Weber, but as Veblen sees it, this "spiritual attitude given by this training in reasoning . . . from pecuniary premises to pecuniary conclusions, is necessarily conservative." This pecuniary mindset further discredits itself in being "unable to take a sceptical attitude toward these postulates or toward the institutions in which these postulates are embodied" (*TBE,* 320). Veblen, as we have seen, praises skepticism (which, he contends, is the "beginning of science" [*HLA* 181]); by denying businessmen the capacity for skepticism, he insinuates they are trapped in institutional thinking. Unlike industry, which continually moves forward, business lags behind; businessmen therefore are "necessarily conservative."

"Business enterprise is an individual matter, not a collective one," declares Veblen (*TBE,* 300). He believes that the individualism of businessmen demonstrates base self-interest, but to Dreiser in the *Trilogy,* it means heroic independence. Furthermore, Dreiser pointedly disagrees with Veblen's claim about the necessarily conservative nature of businessmen. He glamorizes the businessman's individualism, identifying Cowperwood as an avant-garde agent of revolution, much as he romanticizes his own intransigence as a cultural critic.

Dreiser's portrayal of Cowperwood suggests that he must have engaged in some creative misreading of Herbert Spencer, whose influence on him has been ably documented by Ellen Moers, Louis Zanine, Philip

---

of the spiritual truths of law and custom as the circumstances will permit. The realities with which the technicians are occupied are of another order of actuality" (*AO,* 263).

Gerber, Ronald E. Martin, Richard Lehan, and Nancy Warner Barrineau, among others. What William James refers to as Spencer's "scandalous vagueness" made the ideas of the famous Social Darwinist particularly easy for Dreiser to misconstrue. As Richard Hofstadter notes, "the inconsistencies and ambiguities of [Spencer's] system gave rise to a host of Spencer exegesists, among whom the most tireless and sympathetic was Spencer himself."[18]

I do not wish to minimize the importance of Spencerian thought to Dreiser's *Trilogy of Desire*. Some of its operational concepts, particularly the emphasis on "force," clearly derive from Spencer, as do such comments as "the money system of the United States was only then beginning slowly to emerge from something approximating chaos to something more nearly approaching order" (*F,* 34). But while a young Dreiser, writing a 1897 column for the popular music magazine *Ev'ry Month,* could praise Spencer for his "generalship of the mind" and delight that Spencer "has bound the world of knowledge in one," the more mature narrative voice of the *Trilogy* tends to resist, and even to critique, Spencer's confidence in an intelligible universe.[19] For instance the narrator of *The Financier* simultaneously mocks the jury that decides Cowperwood guilty of four counts (not because they have evidence, but because they don't understand the charges) and Spencerian certainty by quipping that "Men in a jury-room, like those scientifically demonstrated atoms of a crystal which scientists and philosophers love to speculate upon, like finally to arrange themselves into an orderly and artistic whole. . . . some vast subtlety that loves order" (*F,* 324-25). That sounds like a parody of Spencer, whose theory of evolution may be summed up, in his own words, as "a change from a less coherent form to a more coherent form" and "from the uniform to the multiform."[20]

Dreiser's debt to Spencer is important, but overstating it occludes substantial differences. Dreiser considered human beings to be

---

18. Moers, *Two Dreisers;* Zanine, *Mechanism and Mysticism;* Philip L. Gerber, "*Jennie Gerhardt:* A Spencerian Tragedy"; Martin, *American Literature;* Richard Lehan, "The City, the Self, and the Modes of Narrative Discourse"; Nancy Warner Barrineau, introduction to *Theodore Dreiser's Ev'ry Month,* xxviii–xxix, xxxiii; James, *The Principles of Psychology,* 1:149 n; Hofstadter, *Social Darwinism,* 41. See also Kevin R. McNamara, "The Ames of the Good Society: *Sister Carrie* and Social Engineering," 223; and Christopher G. Katope, "*Sister Carrie* and Spencer's *First Principles.*"

19. *Theodore Dreiser's Ev'ry Month,* 241, 242.

20. Herbert Spencer, *First Principles,* 359.

considerably less evolved than Spencer did. In *The Titan,* for instance, he reduces "the ultimate dreams of a city or state or nation" to "the grovelings and wallowings of a democracy slowly, blindly trying to stagger to its feet" (440). Or consider the famous passage in *Sister Carrie* where Dreiser describes "[o]ur civilization" as "still in a middle stage, scarcely beast, ... scarcely human" (74). A second fundamental difference in viewpoints—one that has direct relevance to the *Trilogy*—becomes clear when we consider the claim of the most famous American proponent of Spencer, William Graham Sumner, that "[c]apital is only formed by self-denial." It is difficult to imagine a view of capital formation more sharply divergent from that which Dreiser portrays in the character of Cowperwood, who far from denying, seeks ever to satisfy himself.[21]

As David W. Noble argued some years ago, we must resist the "temptation to accept Dreiser's avowed debt to ... Spencer" because in many respects he "rejected the heart if not the details of Spencer's philosophy."[22] Dreiser's commitment to cultural criticism, however unfashionable a form it takes in the *Trilogy,* accounts for his fundamental difference from Spencerian ideology. The sociopolitical views of the English Social Darwinist were fundamentally conservative, working to ensure the maintenance of the status quo.[23] In the *Trilogy,* Cowperwood's "*laissez-faire*" attitude (*F,* 268)—which on the surface sounds like straight Spencer—becomes the means of subverting, not upholding, the moral status quo. Thus Cowperwood manifests many traits that make him an anti-Spencerian figure.

21. William Graham Sumner, *What Social Classes Owe to Each Other,* 67. In "Dreiser and Veblen," Noble argues that Veblen finally held on to more of the Spencerian idea of progress and that Dreiser was the truer Darwinian. Although Veblen claims to have written "Some Neglected Points in the Theory of Socialism" "in the spirit of the disciple" of Spencer, he contemptuously dismisses the notion that "even the evolutionary process ... is infused with a preternatural, beneficent trend; so that 'evolution' is conceived to mean amelioration or 'improvement' " (*POS,* 387, 55). For Veblen's views on Spencer, see also *POS,* 191–92 n 7, 260 n 4. For a contrasting view, see E. Anton Eff, "History of Thought as Ceremonial Genealogy: The Neglected Influence of Herbert Spencer on Thorstein Veblen."

22. Noble, "Dreiser and Veblen," 147, 148. Similarly, in *American Literature,* 216, 233, Martin notes that Dreiser "misunderstood Spencer" on crucial points, generally "modify[ing] ... beyond recognition" Spencer's ideas.

23. Initially, Spencer's views seemed revolutionary because he lionized science in general and evolutionary theory in particular. In a famous incident at Yale, William Graham Sumner fought with the college president, Noah Porter, for the right to use Spencer's *Study of Sociology* in his sociology class. Nevertheless, the ultimate triumph of Spencerian thought in the United States owes much to the conservative (often reactionary) nature of his social and political views.

The Spencerian implications of the most famous passage in the *Trilogy*, where a ten-year-old Cowperwood watches a lobster systematically devour a squid in a fish tank, have been duly noted. This graphic episode surely illustrates what Spencer called the survival of the fittest, and Cowperwood abstracts from it the quintessentially Social Darwinist lesson that "[t]hings lived on each other," including humans on one another (*F,* 8). But equally significant is the revelation that, even as a youth, Cowperwood rejects conventional moral explanations. When "His mother told him the story of Adam and Eve . . . he didn't believe it," and the lesson of the fish tank "cleared things up considerably" for Cowperwood (*F,* 7). He substitutes for the inflexibility of received moral laws his own protean code of "financial morality": "Morality varied . . . with conditions" (*F,* 133, 134).[24] What intrigues Dreiser about Cowperwood's character is this contempt for conventional thinking and willingness to defy it.

Cowperwood is Dreiser's most complete fictional transposition of his own ideal persona as a cultural critic. Especially in the presentation of Cowperwood's moral anarchism, the perspectives of author and character converge.[25] Through Cowperwood, Dreiser continues his battle against the agents of conventional morality that he felt attempted to restrict his work. Throughout the *Trilogy,* Dreiser discredits characters who represent personality types that he felt had been hostile to his work, from the "jealous," "anemic, dissatisfied" girl who sends notes to Edward Malia Butler and Lillian Cowperwood revealing Aileen and Frank's affair; to Lillian herself, with her "conventional mind," "oyster-like in its functioning"; to Chicago's Mayor Sluss ("What is to be done with such a ragbag, moralistic ass . . . ?") (*F,* 186, 216; *T,* 297). The demonization of Cowperwood by Chicagoans throughout *The Titan* is par for the course: Dreiser believed that moral revolutionaries (such as he fancied himself) were inevitably excoriated. The selection of

---

24. The phrase "survival of the fittest" originated with Spencer, not Darwin. In "*The Financier:* Dreiser's Marriage of Heaven and Hell," Stephen Brennan posits that "Dreiser does not really believe life is organized on the basis of animal struggle" (63)—which implies Dreiser's distancing himself from, if not outright rejecting, Herbert Spencer. Brennan illustrated Dreiser's extensive use of religious imagery, but in doing so occludes one of its most salient features: Cowperwood's amoralism.

25. In *In Visible Light,* Shloss begins with a similar premise—Dreiser sees Cowperwood as the heir of Eugene Witla, and thus assigns to the financier exceptional powers of observation—but concludes that Dreiser ultimately exposes the duplicity of commercial achievement (134–35).

Charles T. Yerkes as his primary historical model for the *Trilogy* is again telling: Chicagoans launched a popular crusade against Yerkes's attempt to renew his traction franchise that quickly led to the establishment of a Citizens' Committee of One Hundred to fight him.[26] Having one hundred well-connected enemies opposing Cowperwood provides Dreiser with an opportunity to construct an idealized story of the outsider as successful agent.

Dreiser's view of the financier as a morally revolutionary force commits him to the special pleading that characterizes the *Trilogy*. The narrator of *The Financier* explains that Cowperwood's arrogant and antisocial "I satisfy myself" motto signifies his "intellectual and social nobility" (*F,* 121). Especially useful to Dreiser—for instance, when it comes to defending sexual "varietism," a practice at which Cowperwood excels—the financier acts, as Eugene Witla of *The "Genius"* tries to, like "a law unto himself" (*T,* 121). Dreiser consistently associates lawlessness with a high order of pecuniary success; as he says in a 1912 interview, "Such men as Rockefeller, H. H. Rogers, Jay Gould, William H. Vanderbilt, E. H. Harriman and perhaps Russell Sage. . . . knew no law and they would smile with contempt on any one who did."[27] Not surprisingly, the 1920 essay "The American Financier" contains one of Dreiser's sharpest condemnations of the "so-called laws and prophets" of such concern in his cultural criticism:

> An interviewer once . . . raised the question as to whether the American financial type, then so abundant and powerful, had ethically the right to be as it was or do as it was doing. . . . My answer was, and I still see no reason for changing it, that, in spite of all the so-called laws and prophets, there is apparently in Nature no such thing as the right to do or the right not to do, if you reach the place where the significance of the social chain in which you find yourself is not satisfactory. (*HRDD,* 87)

Throughout the *Trilogy,* Dreiser keeps insisting on Cowperwood's transcendence of institutional restraints. For instance, toward the end of *The Titan,* the narrator asks, "How could the ordinary rules of life or the accustomed paths of men be expected to control" Cowperwood? (478). In fact, the tycoon "knew no law except such as might be imposed upon him by his lack of ability to think" (*T,* 121).

26. Roberts, "Portrait," 352, 356.
27. "Theodore Dreiser Now Turns to High Finance," 197. Dreiser also praises Vanderbilt, Cooke, Rockefeller, and other magnates in *Dreiser Looks at Russia,* 153.

The 1912 edition of *The Financier* insists even more emphatically on the superiority of Cowperwood's lawlessness. The narrator remarks with approval that "above all, [Cowperwood] was no moralist. . . . He saw no morals anywhere" (432). Early in this first edition of the novel, Dreiser correlates his character's amoralism with the work Cowperwood does. Commenting upon the ephemeral nature of honor, the narrator remarks, "You couldn't be generous or kind in times of stress. Look at the conditions on the stock exchange," as if the latter somehow explains the former. The stock market remains a (favorably weighted) emblem of amorality when the narrator continues on to note, "Here men came down to the basic facts of life—the necessity of self care and protection. There was no talk, or very little there, of honor" (102). Tellingly, the narrator then directly swipes at religious authority as "a lot of visionary speculations which had no basis in fact" (103). Having discounted the truthfulness of Christianity, the narrator proceeds to affirm that Cowperwood "preferred to substitute the reality for the seeming" (103). That inversion, strange in that Cowperwood typically prefers the immaterial, affirms the machinations of the stock market as more "real" than the illusions purveyed by conventional morality.

When Veblen describes the businessman as "the only large self-directing economic factor" in modern society, he takes that fact as confirming the businessman's destructive individualism and hostility to the common good (*TBE*, 3). Dreiser, however, interprets Cowperwood's self-direction as indicative of the mighty financier's transcendence of Lilliputian institutions. Both Veblen and Dreiser esteem the revolutionary individuals who they believe defy institutional thinking. The difference is that they locate the defiant mindset in different sectors of the modern economic order: Veblen in industry and Dreiser in business. Consequently, whereas Veblen thinks the engineers and certain social scientists (such as himself) manifest the spirit of skepticism, Dreiser looks longingly toward titanic businessmen like Frank Cowperwood for models of heroic defiance.

Idle Curiosity and the Businessman

Veblen obviously does not think highly of the intellect of your average millionaire: he finds absolutely "nothing recondite about" business methods (*TBE*, 303). The more one understands modern business, he argues, the more facile it all seems. Indeed, a "dispassionate student of the current business traffic, who is not overawed by round numbers, will be more impressed by the ease and simplicity of the manœuvers" of "pecuniary magnates" than by their sophistication (*POS,* 376). Reducing all business activity to the monotonous increase of money, Veblen considers "short-sightedness and lack of insight . . . to be fairly universal traits" of businessmen, and concludes that "[a] man imbued with these business metaphysics" will not likely take up "fine-spun reflection" (*TBE,* 206-7, 232).

Such an attribution might hold for Dreiser's typical male character engaged in business: Charles Drouet, the salesman; George Hurstwood, the manager; or Eugene Witla, the advertising executive. Clyde Griffiths, who is at least nominally engaged in production while working at his uncle's factory (and therefore, according to Veblen's paradigm, advancing industry as well as business) dazzles no one with his brains. From Veblen's perspective, the difference between the lower-level business jobs and the predatory activities of Lieutenants of Finance is one of degree. For Dreiser, however, the difference between his less successful male characters and a Captain of Solvency like Cowperwood is a difference in kind.

According to the business apologist Bruce Barton, success is "eighty-five percent . . . personality," an idea that Dreiser would have seconded. It is, after all, a "personality" in the specifically twentieth-century meaning of that word that Dreiser celebrates in Cowperwood. Despite the modernity of Cowperwood's personality, nostalgia pervades Dreiser's rendering of it, perhaps because, as Robert Wiebe remarks, "As the barons of nineteenth century business retired, their successors appeared to have come from a smaller mold."[28] The exact reason for Dreiser's nostalgia for the large-scale businessman becomes clear in (of all places) *Dreiser Looks at Russia* (1928). While admitting, "I myself [am] ordinarily most sympathetically inclined toward the underdog," Dreiser confesses to feeling sorry that the Soviet Union has

---

28. Barton quoted in T. J. Jackson Lears, "From Salvation to Self-Realization: Advertising and the Therapeutic Roots of the Consumer Culture, 1880-1930"; Robert Wiebe, *Businessmen and Reform: A Study of the Progressive Movement,* 17. In " 'Personality' and the Making of Twentieth-Century Culture," 274-77, Warren Susman contrasts the nineteenth-century model of "character" (associated with the work ethic, duty, and integrity) with the more glamorous twentieth-century model of "personality" (which calls up such descriptions as "fascinating, stunning, attractive, magnetic, glowing, masterful, creative"). Susman connects the shift with such Veblenian notions as increased leisure time, consumption, and attention to appearances (ibid., 280).

"eliminat[ed] . . . the old-time creative or constructive business man" (*DLR,* 75). "Communism or no Communism," Dreiser continues, "it is brain, or cunning, or both—that mysterious something called ability or personality and which same you cannot distribute by law or force—that makes all the difference" (*DLR,* 80). In a 1920 essay, Dreiser elaborates on that difference in personality: "Like the astronomer, the mathematician, the philosopher and the historian, [the financier's] thoughts are more or less remote from the concerns of the ordinary individual" (*HRDD,* 78).

Dreiser once remarked that "Americans, so genuinely clever with machinery and at sports, are, in the matter of economics, perhaps the dullest people in the world" (*TA,* 51). His treatment of the implied reader throughout the *Trilogy* underscores his view of Cowperwood's intellectual superiority. Faced with the yawning dullness of the "uninitiated" (*F,* 94), Dreiser attempts to educate the reader about the intricacies of speculative finance. Thus the narrator of *The Financier* exhorts us: "Imagine yourself by nature versed in the arts of finance, capable of playing with sums of money in the forms of stocks, certificates, bonds, and cash, as the ordinary man plays with checkers or chess. Or, better yet, imagine yourself one of those subtle masters of the mysteries of the higher forms of chess" (99).[29] It seems less that Dreiser is fascinated with the mechanics of the stock market, as might be said of contemporary novelists such as Frank Norris in *The Pit* (1903), than that he doubts his reader's ability to comprehend the subtleties of Cowperwood's business.[30] Particularly in *The Titan,* the narrator loses patience with unbelieving readers, whom he now chastises: "If anyone fancies for a moment that this commercial move on the part of Cowperwood was either hasty or ill-considered he but little appreciates the incisive, apprehensive psychology of the man" (31).

29. The chess image recurs; Dreiser later refers to Cowperwood as "like a master chess player" (*The Stoic,* 213). In "The Financier Himself: Dreiser and C. T. Yerkes," Philip L. Gerber quotes from Dreiser's working notes for the *Trilogy:* "Yerkes should be compared to a man play[ing] chess—the complicated kind with fifteen or twenty opponents. All the moves of each table would be clearly in his mind" (115). Dreiser's image may derive from Ida M. Tarbell's *The History of the Standard Oil Company,* which describes "a brooding, cautious, secretive" John D. Rockefeller "stud[ying], as a player at chess, all the possible combinations which might imperil his supremacy" (27).

30. Dreiser's short story "Phantom Gold" (1927) spotlights the dangers of not understanding financial terminology. Since a poor but greedy character named Queeder "did not in the least understand what was meant by the word option" (259), he provides an easy mark for a well-dressed con artist. The story closes with the financially ruined members of the Queeder family hating each other and the father gone insane.

The intellectual superiority that Dreiser imputes to Cowperwood further supports the moral revolution he is hatching. Thus, for instance, when Dreiser discusses Cowperwood's morally questionable dealings with the Philadelphia city treasurer, Stener, he frames the explanation by noting, "The plan Cowperwood developed after a few days' meditation will be plain enough to any one who knows anything of commercial and financial manipulation, but a dark secret to those who do not" (*F*, 93). The ending frame points again to the inscrutability of finance: "Dark as this transaction may seem to the uninitiated, it will appear quite clear to those who know. . . . It was no different from what subsequently was done with Erie, Standard Oil, Copper, Sugar, Wheat. . . . Cowperwood was one of the first and one of the youngest to see how it could be done" (*F*, 94). Drawing a distinct boundary between the "uninitiated" and those in the know permits Dreiser to stress the intellectual complexity, while downplaying the dubious morality, of Cowperwood's dealings. The net result is that, as H. L. Mencken perceived, Cowperwood emerges as curiously "innocent."[31] In setting his own standards for behavior, Cowperwood does what Dreiser wants the cultural critic to do.

But because Cowperwood remains a capitalist par excellence, most critics have been confused by or indifferent toward Dreiser's portrayal of the financier as an agent of revolutionary change. With few exceptions, readers have rejected Dreiser's insistence that Cowperwood is an extraordinary being, frequently charging that Dreiser just does not understand the nature of business. Donald Pizer is a particularly prominent critic to have claimed that Dreiser "did not realize" a world of "mundane detail and its pervasive emotions of greed, fear, and hate, was incongruous in relation to . . . Cowperwood's more 'refined' instincts," and many others have read the *Trilogy* likewise.[32]

Dreiser knew that his glamorizing of Charles Tyson Yerkes would befuddle readers who expected him to promote a radical social agenda. As he wrote in 1932 to Dorothy Dudley, "I am concluding the last

---

31. Mencken, "Dreiser's Novel," 746. An anonymous 1912 reviewer aptly notes in "Current Fiction" that Cowperwood seems "the standard [in the *Trilogy*]. Other characters are measured thereby" (122).

32. Donald Pizer, *The Novels of Theodore Dreiser*, 169; other rejections of Dreiser's portrayal include Michael Spindler, *American Literature and Social Change: William Dean Howells to Arthur Miller;* Arun Mukherjee, *The Gospel of Wealth in the American Novel: The Rhetoric of Dreiser and Some of His Contemporaries.*

volume of the Trilogy which, I am sure, most of my critics will pounce on as decidedly unsocial and even ridiculous as coming from a man who wants social equity. Nevertheless, I am writing it just that way" (*S*, x). Dreiser's insight into how his books would be read against his intent— and his stubborn determination to write them his way nevertheless— reveals that in his work on Cowperwood, Dreiser's deep and abiding commitment to "social equity" collided with an even more fundamental ideal: the desire to resist moral givens. Cowperwood distills his author's ideal of cultural criticism as intransigence.

Contrary to what most critics have assumed, Dreiser was not interested in anything quotidian or even quantitative in his delineation of Cowperwood. After, as he put it, "look[ing] into the careers of twenty American capitalists," he decided on Charles Tyson Yerkes as his primary model because Dreiser felt Yerkes evoked so much more than the "dull bookkeepers" like Rockefeller.[33] Cowperwood, no dull bookkeeper, seeks something more elusive than money values: he "desired money in order to release its essential content, power" (*S*, 7). Even Veblen admits that "[s]ome large business man may yet rise to the requisite level of intelligence, and may comprehend and unreservedly act upon the fact that the money base line of business traffic at large is thoroughly unstable and may readily be manipulated, and it will be worth going out of one's way to see the phenomenal gains and the picturesque accompaniments [*sic*] of such a man's work" (*TBE*, 208). Dreiser grasps this important idea that money values are unstable— as he puts it, "nothing is so sensitive as money" (*F*, 347)—and in the *Trilogy* he memorializes the "picturesque" consequences of that fact.

In order to see Cowperwood clearly, we must understand his motivation. His quest is not particularly materialistic; rather, it is decidedly philosophical. The narrator of *The Financier* remarks that Cowperwood's "mind, in spite of his outward placidity, was tinged with a great seeking," which accounts for his pursuit not only of wealth, but also of women and art (*F*, 144–45). "We think of egoism and intellectualism as closely confined to the arts," the narrator of *The Financier* reminds us, but "Cowperwood was innately and primarily an egoist and intellectual" (120). Dreiser's insistence on Cowperwood's intellect, in fact,

---

33. Dreiser quoted in F. O. Matthiesen, *Theodore Dreiser*, 129, and in Hussman, *Dreiser and His Fiction*, 71. On the transformation of the historical Charles T. Yerkes into Cowperwood, see the works of Gerber, especially "The Financier Himself."

reads like a point-by-point answer to Veblen's depiction of the businessman. The "depth of [Cowperwood's] nature" is revealed by his inclination toward what Veblen calls finespun reflection: "he was [always] thinking, thinking, thinking" (*F,* 60). In *The Titan,* Cowperwood "often speculated as to what life really was," leading Dreiser to remark that his character "would have become a highly individualistic philosopher," although that calling "would have seemed rather trivial" to the great financier (*T,* 18). Later in *The Titan,* Dreiser justifies the "chronically promiscuous, intellectually uncertain, and philosophically anarchistic" nature of his hero as showing that he "seek[s] the realization of an ideal" (*T,* 186). *The Stoic,* which strikes many readers as inconsistent with the first two volumes of the *Trilogy,* continues to emphasize Cowperwood's bent toward philosophical inquiry. After his death, Berenice Fleming muses that her lover "must know, if he had not when he was here in the flesh, that his worship and constant search for beauty . . . was nothing more than a search for the Divine design behind all forms" (*S,* 327). Although *The Stoic* depicts activities less dramatic or predatory than Cowperwood's exploits in the first two volumes, the philosophical reflections of all three volumes are consistent, as is Dreiser's depiction of his character's pensive nature.[34]

The sort of brooding that Cowperwood engages in resembles what Veblen calls idle curiosity and categorically denies that businessmen can manifest. The disinterestedness that Veblen attributes to idle curiosity puts it fundamentally at odds with the self-interest that he believes typifies businessmen. Furthermore, according to Veblen, idle curiosity is never directed toward a goal (for true speculation is both idle and disinterested) while business is preeminently goal-directed, the end being always differential gain expressed in pecuniary form.

Despite Veblen's assumptions about the mutually exclusive realms of business and idle curiosity, his analysis nevertheless illuminates one of the most salient features of Cowperwood's philosophy: his orientation toward the future. According to Veblen, modern business

34. Many critics cite *The Stoic*'s emphasis on the philosophical dimension of Cowperwood's quest as evidence that the final volume conflicts with *The Financier* and *The Titan.* But the novelist's second wife, Helen, intriguingly suggests in *My Life with Dreiser,* that he "was a mystic, first, last and always" (216). Whether Dreiser was a mystic or not, he emphasizes Cowperwood's philosophical bent throughout the *Trilogy.*
  Comparing the 1912 and 1927 texts, James M. Hutchisson concludes in "The Revision of Theodore Dreiser's *Financier*" that Cowperwood is much more the philosopher in the earlier (and less well-known) version (205).

increasingly hinges less on current output than on what firms define as their " 'presumptive earning-capacity' " (*TBE*, 127). Huge gaps can separate an industry's production (actually, in the present) and a businessman's inflated prediction of its capacity to produce (hypothetically, in the future). Calculations based, "not [on a concern's] past or actual earning-capacity, but [on] its presumptive future earning-capacity" simply "turn about imagined future events" (*TBE*, 153, 154). But for Veblen, this mode of calculation is pure fantasy. He counters the businessman's self-deluded prediction of the future, again, with the lessons of industry, where there can never be "a final term in any direction. . . . The ramification of industrial dependence is without limits" (*TBE*, 187-88). Veblen's scheme no more allows a predetermined future in industry than Darwin's does an end to evolution; like his predecessor in biology, Veblen's social and economic theory is thoroughly antiteleological.

Few fictional businessmen in American literature receive so romanticized a treatment as Cowperwood. In sharp contrast with one who does–Fitzgerald's Jay Gatsby, who looks always to the past—the philosophy of Dreiser's character is practically defined by his futurity.[35] It is in the imaginative space of future prediction defined by Veblen as central to businessmen's thinking that Cowperwood's mind operates, and this orientation explains his financial success. His "implicit faith in the future of the street railway" helps him to keep succeeding throughout his life (*F*, 53). In *The Titan,* Cowperwood's ability to look at Chicago and see not a rough young city but the "future Chicago" accounts for his victories (*T*, 23). Taking over the Chicago gas industry is, like Cowperwood's railway dealings in Philadelphia, a fortunate bet on "future prospects" (*T*, 93). Not surprisingly for a man who "kept his face and thoughts to the future," one of Cowperwood's greatest hardships during his incarceration in Philadelphia is not being able to "attend to any of the things he ought to be attending to—his business affairs, his future" (*F*, 217, 394). Probably one reason the philandering Cowperwood never divorces his second wife, Aileen, is because her steadfast belief in his future touches him to the core (*F*, 367).

35. In F. Scott Fitzgerald's *The Great Gatsby*, the title character remarks, "Can't repeat the past? . . . Why of course you can!" Gatsby wants a censored and sanitized version of the past: "He wanted nothing less of Daisy than that she should go to Tom and say: 'I never loved you.' After she had obliterated four years with that sentence they [Gatsby and Daisy] could. . . . go back to Louisville and be married from her house—just as if it were five years ago" (111).

Cowperwood, who was, unlike Veblen, quite a teleologist, imagines that he knows how everything will turn out. The 1912 *Financier* even grants him "clairvoyance. He could fairly see and feel in advance what was coming" (106). Predicting how the Chicago railway situation will look several years hence, he declares, "I can see it coming" (*T,* 161). The belief that he knows the future helps Cowperwood to sustain his sublimely arrogant self-image. When the panic following the 1871 Chicago fire makes the future look "hopeless" to his rival, Edward Malia Butler, Cowperwood feels "hope" (*F,* 219). No matter what happens, "somehow he always believed in his star" (*F,* 240).

Dreiser's own attempts in *The Trilogy of Desire* to foretell the future with all the confidence of a Cowperwood lead to some curious results. One of the most bizarre examples occurs in the epilogue to *The Financier,* where the narrator takes up "THE MAGIC CRYSTAL" and imagines himself reading a crystal ball (448). The last word in *The Financier,* appropriately enough, is "end" (448). Insofar as Dreiser's tale of financial drama draws from historical personages and events, the narrative's end *is* foretold. But some strange fictionalizing comments also predict the future, especially in the volume most informed by a cyclical notion of time, *The Titan,* which emphasizes Cowperwood's rebirth and renewal. The narrator repeatedly interrupts the swift pace of events in *The Titan* with comments told as if from the end of time such as, "If it had not been for Cowperwood's eventual financial triumph over all opposition. . . ." (100). Numerous criticisms of Aileen by *The Titan*'s narrator likewise make Cowperwood's emotional defection from her seem a foregone conclusion. Given the significant number of predetermined outcomes, owing to the fact that Dreiser worked closely from the historical record, it is curious that he had so much trouble finishing the third volume.

Cowperwood's philosophical bent again sheds light on why Dreiser had trouble letting this character go. Throughout the *Trilogy of Desire,* Dreiser, like his hero, wars with himself about whether or not there is a discernible pattern to existence (and if not then any notion of foretelling the future is clearly moot). The lobster and squid episode begins Dreiser's projection of his own philosophical quandaries onto his character, much as Cowperwood projects human consequence onto the animals (referring to each one as "he," not "it"). The financier's epiphany, it should be noted, comes in response to his "forever pondering, pondering . . . how this thing he had come into—this life—was organized" (*F,* 7). Once the lobster devours the squid, young Frank feels he has an answer to "that riddle which had been worrying him so much in the past: 'How is life organized?' " (*F,* 8).[36] As a ten-year-old, Cowperwood achieves the serene confidence of a social theorist who has discerned his own place in the universe. Dreiser repeats the phrase again after Cowperwood witnesses the brutality attending the birth of his first child, which startles him briefly into conventionality: "That old conviction of tragedy underlying the surface of things. . . . There was a good deal to this home idea, after all. That was the way life was organized" (*F,* 58).

Dreiser repeats the phrase because he agonized over the same thing. Early in *Dawn,* he describes his mother as a "sweet, dreamy woman, who did not know how life was organized" and himself as wishing he knew "a way by which life might be better organized" (10, 22). In the *Trilogy,* Dreiser both questions and endorses Cowperwood's belief that he has figured out the answer; sometimes he does both in the same passage. During a section of free indirect discourse, for instance, the narrator makes us privy to Cowperwood's mind: "Life was a dark, insoluble mystery, but whatever it was, strength and weakness were its two constituents" (*F,* 241). The strength and weakness idea is right out of the fish tank; but how can such a hard-boiled and radically simplified idea coexist with a declaration of "insoluble mystery"? Equating this mystery with Spencer's "unknowable" does not resolve the contradiction, for that murky premise of *First Principles* concedes to the same religious beliefs that Dreiser's *Trilogy* assails.[37] The lobster and squid, we are told, "cleared things up considerably intellectually" for Cowperwood (*F,* 7), but elsewhere for instance, the narrator of *The Financier* generalizes, "We all know that life is unsolvable—we who think. The remainder imagine a vain thing, and are full of sound and fury signifying nothing" (*F,* 187), and comments that "Life to him [Mollenhauer, one of the Philadelphia "Big Three"], as to every other man of large practical knowledge and insight, was an inexplicable

---

36. An origin for the lobster and squid passage is in Dreiser's "A Lesson from the Aquarium." Gary Scharnhorst suggests in "Reconstructing Here Also: On the Later Poetry of Charlotte Perkins Gilman," that Gilman may have had Cowperwood in mind when she writes in "The Oyster and the Starfish" of an anthropomorphic scavenger that devours shellfish (259–60).

37. Hofstadter, for instance, in *Social Darwinism,* describes Spencer's "Unknowable" as his "important concession to religion," without which his positivistic views would probably have been "totally unacceptable in America" (37).

tangle" (*F,* 200). In such declarations, Dreiser sounds more like Darwin in the *Origin of Species,* reminding arrogant humans how very little we know—or like Veblen insisting that epistemological certainty is a sham—than like Spencer in his certainty that he knows the answers and that they reflect comfortably on mankind (particularly on Anglo-Saxon males).[38]

Not even the financier is fully convinced by his certainty. Something unstable in his "dynamic and seemingly unterrified approach to life" (*S,* 205) keeps surfacing, particularly in *The Financier.* During Cowperwood's incarceration, Dreiser discloses that "The strongest have their hours of depression" (*F,* 396)—which is rather like saying that the strongest have their hours of weakness, that the lobster has moments of feeling squidlike. Still affirming Cowperwood's "strange self-confidence," the narrator distances himself from his character:

> It would be too much to say that Cowperwood's mind was of the first order. It was subtle enough. . . . It was a powerful mind, . . . but it was not sufficiently disinterested to search the ultimate dark. He realized, in a way, what the great astronomers, sociologists, philosophers, chemists, physicists, and physiologists were meditating; but he could not be sure in his own mind that, whatever it was, it was important for him. . . . the call of his own soul was in another direction. His business was to make money—to organize something . . . , or, better yet, save the organization he had begun. (*F,* 396)

Self-interest, the factor Veblen claims makes idle curiosity alien to businessmen, may inhibit Cowperwood's inquiry after all. Such qualifications enrich Dreiser's characterization of the financier and raise a number of questions. Does Cowperwood understand how life is

38. One of Darwin's most striking images of human ignorance occurs in *The Origin of Species by Means of Natural Selection, or The Preservation of Favored Races in the Struggle for Life:* "I look at the natural geological record, as a history of the world imperfectly kept, and written in a changing dialect; of this history we possess the last volume alone, relating only to two or three countries. Of this volume, only here and there a short chapter has been preserved; and of each page, only here and there a few lines. Each word of the slowly-changing language, in which the history is supposed to be written, being more or less different in the interrupted succession of chapters, may represent the apparently abruptly changed forms of life, entombed in our consecutive, but widely separated formations" (316). As Darwin says, "It is so easy to hide our ignorance under such expressions as the 'plan of creation,' 'unity of design,' &c., and to think that we give an explanation when we only restate a fact" (ibid., 453). Of course such phrases are legion in Spencer.

organized? Is there a connection between his belief that he knows how life is organized and his need to make his life's work (as the last quoted passage mentions twice) organization? I do not think Dreiser's qualifications undermine Cowperwood but that the author's deep identification with his character—particularly for his moral anarchism, self-confidence, and ability to realize his goals, all traits that Veblen associates with engineers rather than financiers—leads Dreiser to project his deepest philosophical quandaries onto Cowperwood.

When Dreiser distances himself from his hero during the incarceration episode, he describes Cowperwood's self-confidence as "strange" (*F,* 396). But Cowperwood is really no stranger to the "sensation of strangeness and defeat" that comes over him; similar moments of vulnerability surface, especially in *The Financier* (329). What surprises us is how quickly Cowperwood bounces back, how soon he "regain[s] completely his critical attitude, his cool, intellectual poise" after each setback (*F,* 329). When Cowperwood faces unanswerables, as when he gazes at the stars from prison, he is able to "sh[ake] these moods off" and recover "a sense of grandeur, largely in relation to himself" (*F,* 418) with an ease that Dreiser can only envy. Cowperwood cons himself into believing he has the answers—and thus his appeal for Dreiser, even more than for his pecuniary and amorous successes.[39]

Dreiser concurs, finally, with Cowperwood that "Life and character must really get somewhere to be admirable," and in doing so breaks with Veblen's preference for the "idleness" of curiosity (*T,* 108). Getting somewhere may be easier if one is not distracted by glimpses of the ultimate dark, and so Dreiser admires the instrumentality of Cowperwood's philosophy. Cowperwood's self-confidence and his belief that he knows how life is organized permit him to defy conventions. If the admirable traits of a character result in tangible accomplishments, and if Cowperwood's certainty facilitates his achievements, then from a pragmatic standpoint it becomes irrelevant whether he is deluding himself. How is life organized? Consider the end of *The Stoic:* in the chapters following Cowperwood's death, his empire faces many setbacks, yet

---

39. In *Gold Standard,* Michaels sees *The Financier* demonstrating that "the business crisis, understood by Dreiser as an essentially natural phenomenon, cannot be mastered or even predicted by any system of thought, any account of life's 'organization'— theological, scientific, even economic" (77). But Dreiser admires Cowperwood's certainty (even if it is self-deluded) that he has the answers.

his organizational power has outlasted his mortal body.[40] As Lord Stane says during Cowperwood's final illness, "Cowperwood is too great an organizer to found any vast enterprise on the indispensability of any one man" (*S*, 271). Here fact and fiction merge: the Chicago railway and London underground system both show that life is organized by people like Cowperwood-Yerkes.

Hence the contradictory statements about whether Cowperwood does, or does not, know how life is organized are not simply complications within Dreiser's characterization or signs of the author's ambivalence toward business. As Ellen Moers argues, there are in fact "two Dreisers," a startlingly modern one and a backward-looking Victorian. The modern side of Dreiser creates a Cowperwood who defies conventional explanations of the world, while the Victorian side manifests itself in the longing for a master system with coherent rules that account for how life is organized. Because Dreiser wants desperately to know how it all works, and because he doubts that he ever can, he reveres Cowperwood.

Making Believe

Cowperwood's self-delusions would not surprise Veblen, who felt that business layered pretence on top of pretence. Indeed, business exemplifies the ceremonialism he detests: "The realities of the business world are money-values; that is to say matters of make-believe which have the sanction of law and custom and are upheld by the police in case of need" (*AO*, 108). Financiers, as we have seen, characteristically make money by exploiting the gray area between make-believe and reality, "induc[ing] a discrepancy between the putative and the actual" (*TBE*, 156). The complicity of law and custom adds another level of untruth, while the anachronistic theories of orthodox economists further promote the mass delusion: "there is a naïve, unquestioning

---

40. Consider also the explanation in *Sister Carrie* of how a man's business can outlast him. Dreiser begins with a biological analogy, only to conclude that the intelligent rich man can finally beat biology: "A fortune, like a man, is an organism which draws to itself other minds and other strength. . . . Beside the young minds drawn to it by salaries, it becomes allied with young forces, which make for its existence even when the strength and wisdom of the founder are fading. . . . This removes it at once beyond the special care of the founder." Thus "The man wanes, . . . and the fortune . . . continues," as we see with Cowperwood (305).

persuasion abroad among the body of the people to the effect that, in some occult way, the material interests of the populace coincide with the pecuniary interests of those business men" (*TBE*, 286). Veblen considers this notion a "picturesque hallucination" (*VI*, 133). Besides disparaging the "hallucinations" that businessmen manufacture to delude the public, he attends to their self-delusions. For instance, the "acquisition of property" is taken "to mean production of wealth; so that a business man is looked upon as the putative producer of whatever wealth he acquires" (*TBE*, 291). Such shams reveal the ultimate reality of business, Veblen claims, as "fictitious" (*TBE*, 103, 104).

That fictitiousness causes Dreiser to identify with Cowperwood. In *The Trilogy of Desire* the financier remains "true to his policy of make-believe" (*F*, 383). Similarly, Cowperwood never has to pay back the money he stole from the city of Philadelphia due to "some hocus-pocus" (*F*, 415).[41] Cowperwood's first enterprise of his own, Cowperwood & Co., has as its basis an illusion: "The company was fiction purely" (*F*, 68). Similarly, once public opinion in Philadelphia solidifies against him, Cowperwood establishes a silent partnership with Stephen Wingate, another fictitious alliance, for the financier choreographs his supposed partner's every move (*F*, 350). In *The Titan*, Cowperwood breaks into Chicago with more illusions, installing his own dummy directors in gas companies. These acts of legerdemain confirm the illusory nature of Cowperwood's work, in which "fluctuations" in value are not real but "manufactured," and in which "creating a fictitious demand" by "much fictitious buying" is common practice (*F*, 94, 141). What Cowperwood primarily deals in, after all, is what Veblen describes as capital "subdivided into convenient imaginary shares" (*TBE*, 155).

Veblen extrapolates from the idea of the illusory nature of business to an extreme and radical conclusion. He considers one of the greatest conceits of the business community the notion that the money values it measures actually signify something concrete and stable (*TBE*, 105-6). Veblen deduces from his premise that the "base line" of business is chimerical "money value," not anything tangible, the conclusion that business dissociates itself from "material efficiency" (*TBE*, 105). This denial of any material, substantial basis for business enterprise

---

41. Dreiser's primary model for Cowperwood, Charles T. Yerkes, affirmed the market value of illusions: "The secret of success in my business is to buy old junk, fix it up a little and unload it upon other fellows" (quoted in Matthew Josephson, *The Robber Barons*, 386).

forms one of the pillars of Veblen's economic theory. It follows from his distinction between industry and business, and coincides with Veblen's allegations about the "make-believe" quality of the latter. As he expresses this central paradox, the "substantial foundation" of the modern corporation is its "immaterial assets" (*TBE,* 143).

Insisting upon the immaterial basis of business permits Veblen to strike precisely on the quantitative ground where it thrives. "[T]he substance must not be allowed to stand in the way of the shadow," he remarks ironically (*HLA,* 106). When he argues that "[i]n the business world the price of things is a more substantial fact than the things themselves" (*AO,* 89), he means quite literally that cost (for the consumer) and profit (for the businessman) take precedence over the actual objects bought and sold. Because "no hard and fast relation" connects pecuniary accountancy to industry, Veblen can conclude that business is "shifty" (*TBE,* 131, 132). Business is "shifty" both because its relationship to industry fluctuates and because it is dishonest, deceptive. For Veblen, the immaterial basis of modern business defines its essence and proves its inauthenticity.

Dreiser also constructs *The Trilogy of Desire* on the idea of business as immaterial, but the absence of a concrete basis for business signifies for him transcendence and subtlety, not fraudulence and make-believe. Thus he describes Cowperwood's work: "His business . . . was with the material facts of life, or rather, with those third- and fourth-degree theorems and syllogisms which control material things and so represent wealth" (T 18). From his youth, Cowperwood comprehends the "vast ramifications of finance"; "[t]he abstrusities of the stock exchange were as his A B C's to him" (*F,* 11, 92). Dreiser seeks in the *Trilogy,* as in his other novels, to depict what lies behind the material world. As H. L. Mencken so aptly notes, "the thing [Dreiser] exposes is not the empty event and act, but the endless mystery out of which it springs." The immaterial basis of business represents endless mystery to Dreiser, and the entire *Trilogy* spins out the idea that "this material seeming in which we dwell is itself an illusion" (*F,* 447).[42]

Successful operators like Cowperwood employ what Veblen describes as "intangible assets" that give one businessman a "differential advantag[e]" over another (*POS,* 360–61). For instance, an able

42. H. L. Mencken, "The Dreiser Bugaboo," 770. Dreiser's preoccupation with immateriality in the *Trilogy* extends also to sexual attraction. What Aileen desires in Cowperwood is "not his body—great passion is never that," and his attraction to Berenice Fleming is "nonmaterial" (*Financier,* 123; *Stoic,* 146).

businessman's capital incorporates "psychological phenomena" such as "good will, fashion, customs, prestige, effrontery, personal credit" (*POS*, 311). Cowperwood enjoys many unquantifiable assets, including the penultimate item in Veblen's list (Chicago newspapers describe Cowperwood as succeeding in London "by reason of his wealth, cunning, and general effrontery" [*S*, 133]), but the asset that Veblen calls "good will" best illustrates the paradox of immaterial substance. Citing Poor Richard's "honesty is the best policy" as an example, Veblen explains that goodwill originated as an "immaterial by-product of the concern's conduct of business," "impl[ying] a kindly sentiment of trust and esteem" (*POS*, 366, 363). But goodwill has degenerated under modern capitalism so that it is no longer "serviceable to the community, but only to their owners" (*TBE*, 139).

Cowperwood, knowing that "[n]othing was so precious as reputation and standing" (*F*, 142), grasps at an early age the role goodwill plays in business. Even at his first job working for the grain and commission brokers, Waterman & Company of Philadelphia, he has the "uncanny faculty" to make others believe in him. Cowperwood "strik[es] balances for his employer, pick[s] up odd lots of things they needed, solicit[s] new customers, break[s] gluts by disposing of odd lots in unexpected quarters. Indeed the Watermans were astonished at his facility. . . . for getting appreciative hearings, making friends, being introduced into new realms. New life began to flow through the old channels of the Waterman Company" (*F*, 28). Dreiser's metaphor, drawing upon the physical realm to explain the immaterial, suggests goodwill is the lifeblood of an enterprise.[43]

The pecuniary value of this intangible asset becomes evident in *The Financier* when Cowperwood's failure follows the city treasury scandal. The temporary decline in Cowperwood's goodwill pulls the value of stocks down by three points (*F*, 220). Inversely, the able financier can use goodwill to generate capital. In *The Titan,* for example, Cowperwood finds the banks closed to him when he is desperately in need of capital to finance his planned elevated lines. His credit limited,

43. The 1912 edition is again even more explicit. Early in his career, while pursuing a block of government bonds to help finance the Civil War, Cowperwood acknowledges that "money was the first thing to have," but goodwill must back it: "Then the reputation of handling it [money] wisely would treble, quadruple, aye, increase its significance a hundred and a thousand fold. First you secured the money. Then you secured the reputation. The two things were like two legs on which you walked. Then your mere word was as good as money, or better" (1912 *F*, 125).

Cowperwood reckons on getting what he needs by extending his goodwill. He plans a gift to the University of Chicago, calculating that "On such a repute (the ability to give a $300,000 telescope out of hand to be known as the Cowperwood telescope), he could undoubtedly raise money in London, New York, and elsewhere for his Chicago enterprise. The whole world would know him in a day" (*T,* 337). Cashing in on the "unique news value" of his gift, Cowperwood presents himself as a "public benefactor and patron of science" (*T,* 338-39). The result: "he could undoubtedly raise money" previously unavailable to him (*T,* 337).

Veblen notes that donations to universities are common "investment[s] in good fame" that accrue "pecuniary respectability" to the donor (*HLA,* 156).[44] Cowperwood's gift of the telescope to the University of Chicago illustrates that point as well as the immaterial basis of business: his material assets do not grow—in fact one would expect them to have shrunk by $300,000—but his immaterial assets skyrocket. Furthermore, Cowperwood gives the university tangible goods (land, building, telescope) to create intangibles ("news value" and "repute"), which, in turn, generate the money to purchase the tangible "steel, . . . labor, and equipment" (*T,* 338, 337, 331) for his elevated lines. Thus the material and immaterial realms can cross over, in alarming defiance of mathematical and physical laws. A familiar example of this phenomena that Veblen provides is advertising, which invests tangible assets (such as paper) to produce an intangible asset (the ad) which in turn increases sales and, thereby, augments tangible assets (*POS,* 367-68).

Veblen claims the categories of tangible and intangible are in fact "seldom clearly distinguishable" in modern business, and he finds this blurring even more outrageous than the value of intangibles (*TBE,* 155). Consider, for instance, his analysis of the capitalization of U.S. Steel. The merger was funded by the " 'good-will' " of Andrew Carnegie and others, "But good-will on this higher level of business enterprise has a certain character of inexhaustibility, so that its use and capitalization in one corporation need not, and indeed does not, hinder or diminish the extent to which it may be used and capitalized in any other corporation"

---

44. For similar reasons, John D. Rockefeller insists in *Random Reminiscences of Men and Events* that "it is not capital and 'plants' and the strictly material things which make up a business, but the character of the men behind these things, their personalities, and their abilities; these are the essentials to be reckoned with" (94-95). Veblen would identify conspicuous consumption, which "induces the American millionaire to found colleges, hospitals and museums," as motivating Cowperwood (*TLC,* 91).

(*TBE*, 173).⁴⁵ Under the aegis of goodwill, capital defies the laws of physics, residing in more than one place at a given time.

Veblen even more radically loosens business from the material world, in a passage remarkable enough to warrant quoting at length:

> The case is analogous, though scarcely similar, to that of the workmanlike or artistic skill of a handicraftsman, or an artist, which may be embodied in a given product without abating the degree of skill possessed by the workman. Like other good-will, though perhaps in a higher degree of sublimation, it is of a spiritual nature, such that, by virtue of the ubiquity proper to spiritual bodies, the whole of it may undividedly be present in every part of the various structures which it has created. . . . It has also the correlative spiritual attribute that it may imperceptibly and inscrutably withdraw its animating force from any one of its creatures without thereby altering the material circumstances of the corporation. (*TBE*, 173-74)

Veblen borrows the idea of chemical "sublimation," whereby a solid substance is converted by heat into a vapour, to underscore both the immaterial basis of the U. S. Steel transaction and its slippages between the material and immaterial realms. But the "spiritual," "ubiquit[ous]," and "inexhaustible" qualities of goodwill suggest that Veblen also toys with another meaning of sublimation, the "elevation to a higher state or plane of existence."⁴⁶

One of Cowperwood's distinctive financial ploys is to sublimate in the manner analyzed by Veblen above. The trick involves capitalizing on the ability of money to be in several places at once. Cowperwood "knew instinctively what could be done with a given sum of money—how as cash it could be deposited in one place, and yet as credit and the basis of moving checks, used in not one but many other places at the same time" (*F*, 99). Cowperwood's Chicago empire also demonstrates Veblen's point that increases in "business capital . . . swell the volume of business, as counted in terms of price, etc., but they do not directly

---

45. Even the muckraker Gustavus Myers pauses in *History of the Great American Fortunes* (a book that influenced Dreiser's understanding of business) when he comes to the sum of money needed to capitalize U. S. Steel: "And we feel irresistibly constrained to linger upon that billion dollars. The ordinary human mind is capable of much; it can let its exuberant imagination create heavens and hells, enchantments and exorcisms, and it can stretch illusion to realms without limit; but to conceive of a billion dollars, or rather to visualize it, is a task to be forsworn" (596).

46. See *Oxford English Dictionary*, definitions 1 and 5.

swell the volume of industry" (*TBE*, 99). In *The Titan,* Cowperwood extends the capitalization of his Chicago street railways (capitalized in 1886 at six to seven million dollars) tenfold; he inflates the market value of his West Side corporation to "three times the sum for which it could have been built" (*T,* 429); and he gets his North Chicago company "valued at over one hundred thousand dollars more per mile than the sum for which it could actually have been replaced" (*T,* 429).

Because the value of things bears so little relation to their material components, prices seem utterly capricious, as seen especially in the stock market's fluctuations. As the narrator of *The Financier* remarks, "It was useless, as Frank soon found, to try to figure out exactly why stocks rose and fell. Some general reasons there were, of course, . . . but they could not always be depended on" (*F,* 39). Foremost among those "general reasons," the "professional traders were, of course, keen students of psychology" (*F,* 41). Cowperwood, we should recall, has an "incisive, apprehensive psychology" (*T,* 31); that he "knew a great deal about human nature" also serves him well (*F,* 203). Being able to "fit himself in with the odd psychology of almost any individual," Cowperwood approaches potential partners—and potential foes—"at that psychological moment" when they are most likely to join with, or capitulate to, his desires (*T,* 26, 28).

The role played by the psychology of traders in determining stock prices illustrates one of the most immaterial features of modern business. Dreiser explains: "those inexplicable stock panics . . . had so much to do with the temperament of the people, and so little to do with the basic conditions of the country" (*F,* 138). As one of Cowperwood's early mentors puts it, "the rumor that your second cousin's grandmother has a cold" can make or break a market; "It's a most unusual world. . . . No man can explain it" (*F,* 39). Veblen, however, sees banality rather than mystery in the stock market. Fluctuations derive from "variations of confidence on the part of the investors, . . . current belief . . . , forecasts . . . , and on the indeterminable, largely instinctive, shifting movements of public sentiment and apprehension." Thus "under modern conditions the magnitude of the business capital and its mutations from day to day are in great measure a question of folk psychology rather than of material fact" (*TBE*, 149).

Furthermore, according to Veblen (who felt that economic cycles were grossly misunderstood), inflationary and deflationary periods— to say nothing of the more dramatic booms, crises, and busts—are also psychological phenomena. What we describe as an economic "depression" literally results from the mental condition that we describe with the same word—"a malady of the affections" of the businessman (*TBE*, 237). The "primary hardship of a period of depression is a persistent lesion of the affections of the business men," says Veblen, and an economic crisis "rests on emotional grounds" (*TBE*, 239, 238). When businessmen grow depressed, the entire community is at risk.

Representative Man

In one of Dreiser's most blatant assertions of Cowperwood's superiority, he distinguishes the desires of most people from those of his hero. "Few people have the sense of financial individuality strongly developed," the narrator remarks: "They want money, but not for money's sake." Most of us want money for what it will "buy," whereas Cowperwood sees money as only the means to an end. The financier pursues money as a route to what it will "control" and, especially, for what it will "represent," such as "dignity, force, power" (*F*, 182).

Representation occurs when one thing stands for another (or claims to). Representation can be a matter of signification, for instance, when a word, image, or sign evokes something else; it can also be a question of politics, as when an individual or group claims to represent a larger body. Throughout the *Trilogy*, Dreiser insists that Cowperwood's business, his speculative and his combinatory dealings, are instances of representation in the sense of signification—and he seems unconcerned by the fact that Cowperwood represents, politically speaking, no one but himself. In fact, Dreiser uses the idea of finance as a system of representation to set Cowperwood favorably apart from lesser characters. Cowperwood's first employers cannot grasp that finance is a system of elaborate representation; from the Watermans' perspective, "their business significated itself"—nothing else (*F*, 29). Yet for Cowperwood, business signifies far more than itself—it represents immaterialities.

Cowperwood's love of representation, particularly those culturally sanctified forms that we call fine art, is one of the keys to his character and one of the surest indications of how favorably Dreiser reads the immaterial basis of business. "Finance is an art," the narrator declares periodically (*F*, 120) and insinuates throughout the *Trilogy*. Cowperwood, in fact, prefers artistic representation to the putatively real thing represented: "He admired nature, but somehow . . . he fancied one could best grasp it through the personality of some interpreter" (*F,* 60). He buys art until his "collection had become the most important in the West—perhaps in the nation" (*T,* 346).[47] Cowperwood even falls for Berenice Fleming, not because her physical form transfixes him, but when two pictures of her magnetize him (*T,* 316). Fortunately for their relationship, "she fulfilled all the promise of her picture" (*T,* 321).

Cowperwood's artistic approach to finance demonstrates his distinctive notions of representation. When he examines the decrepit tunnels in Chicago, considering their possible uses to him, he looks with the eyes of "a Daumier, a Turner, or a Whistler"; "He forever busied himself with various aspects of the scene quite as a poet might have concerned himself with rocks and rills" (*T,* 157, 156). In fact, Dreiser depicts the three areas in which Cowperwood excels—the acquisition of art, the seduction of women, and the "great art" of finance (*F,* 11)— as representations of each other. Cowperwood's three primary desires are intimately linked in his mind, and their linkage informs the logic of *The Trilogy of Desire.* As the narrator puts it, Cowperwood's "mind, in spite of his outward placidity, was tinged with a great seeking. Wealth, in the beginning, had seemed the only goal, to which had been added the beauty of women. And now art, for art's sake . . . and to the beauty of womanhood he was beginning to see how necessary it was to add the beauty of life—the beauty of material background—how, in fact, the only background for great beauty was great art" (*F,* 144-45). This "great seeking" defines Cowperwood's philosophy and determines his actions, making him an imaginative, if opportunistic, artist in his own right.

Veblen, however, sees business as all "artless routine" and denies that anything done by businessmen can be considered imaginative: "In his capacity as business man he does not go creatively into the work of perfecting mechanical processes and turning the means at hand to

---

47. In "Financier," Gerber notes that the 1910 sale of Yerkes's art collection netted $2,207,866, "the most lucrative sale of its kind to be held" in America (113). On Cowperwood as an artist see Mencken, "Dreiser's Novel," 746; James T. Farrell, "Theodore Dreiser," 155-69; Kenneth Lynn, *The Dream of Success: A Study of the Modern American Imagination,* 42; Charles Child Walcutt, *American Literary Naturalism: A Divided Stream,* 205-6; Gerber, *Theodore Dreiser,* 99-122; Pizer, *Novels,* 151, 169; Hussman, *Dreiser and His Fiction,* 75; Spindler, *American Literature,* 72; Mukherjee, *Gospel of Wealth,* 85; Carl Smith, *Chicago and the American Literary Imagination, 1880-1920,* 71-76; Horwitz, *By the Law of Nature,* 201.

new or larger uses" (*POS*, 376; *TBE*, 44). The work Veblen sees as creative, such as improving technology and enhancing productivity, occurs in the realm of industry, and he insists on "the indispensably creative function of technology" (*AO*, 63). Veblen sees the representation of business as more a political than an aesthetic matter, and finds that business interests have been extremely successful in representing themselves to the public. Take for instance the "Captain of Industry." According to Veblen, in reality "his captaincy is a pecuniary rather than an industrial captaincy," which means that the figure of speech reverses reality (*TLC*, 230). Veblen deconstructs this common representation of the businessman; he claims he is a producer only by "euphemistic metaphor" (*HLA*, 208).

When figurative language occludes reality, Veblen manifests a horror of representation. As we have seen, Veblen finds other social scientists particularly guilty of perpetuating figures of speech that misrepresent reality. Veblen critiques the forms of representation characteristic of orthodox economists, which typically glorify the businessman and mislead the general public. In fact, "the whole 'money economy,' with all the machinery of credit and the rest, disappears in a tissue of metaphors to reappear theoretically expurgated, sterilized, and simplified into a 'refined system of barter' " (*POS*, 250). As Veblen says, "The metaphors [of orthodox economists] are effective, both in their homiletical use and as a labor-saving device"—but much of the labor saved is that necessary for clear thinking (*POS*, 66). Veblen's concern with the deceptive process of representation as a means of signification shades into his critique of political (mis)representation. As he remarks in *Business Enterprise*, "Representative government means, chiefly, representation of business interests" (286). Orthodox economists likewise represent business interests, and misrepresent reality, by their system of representation.

In the preface to *The Theory of Business Enterprise*, Veblen warns readers that his conclusions may seem "unfamiliar" because he assumes "the point of view . . . given by the business man's work,—the aims, motives, and means that condition current business traffic" (xix). The result is a book that may be Veblen's most straightforward and least satirical, although it is surely one of his most critical. His claim to represent the businessman's point of view, however, is certainly ironic.

The question of point of view is of paramount importance also in *The Trilogy of Desire*. First, the perspective of author and main character merge more completely here than in any other Dreiser novel. As he commented to an interviewer in 1912, "In the new novel [*The Financier*], the note of the plot will come from the man," and Dreiser's point of view also comes from Cowperwood.[48] Second and of equal importance, Dreiser insists that Cowperwood's point of view distinguishes the financier from almost everyone around him. Thus the adoption of the businessman's vantage point that Veblen ironically claims for himself in *Business Enterprise,* Dreiser enacts, without the irony, in the *Trilogy.*

A second of Veblen's ironic statements, this one from *Absentee Ownership,* also helps to identify the perspective of Dreiser's *Trilogy of Desire.* "Money-values are the underlying realities of the business man's world, the substance of things hoped for, the evidence of things not seen," Veblen claims (*AO,* 180). *The Trilogy of Desire* also defines business as pursuing invisible and ephemeral forces beyond the substantial world. But whereas Veblen probes the immaterial basis of business in order to expose its unreality, its make-believe status, Dreiser finds the immaterial basis of business a key to philosophical mystery: "Or would you say that this material seeming in which we dwell is itself an illusion?" (*F,* 447). He reveals not only the subtleties of the financial world, but also the strangely fictitious nature of what would seem to be hard reality. Subscribing to Veblen's claim that "[t]he substantial foundation of the industrial corporation is its immaterial assets" (*TBE,* 143), Dreiser turns the judgment around and depicts business as unusual. Dreiser admires Cowperwood for defying convention, treating the law with contempt, and behaving as if institutions do not apply to him. Far from being confined by pedestrian and mundane facts, Cowperwood harnesses the immaterial basis of business and towers above characters who would condemn him. Thus Cowperwood extends into fiction the romanticization of the intransigent cultural critic central to Dreiser's own persona. For Veblen, figures such as Cowperwood form the object of cultural criticism—not a symbol of its potentiality. But both Veblen's view of "business as usual" and Dreiser's view of business as spectacularly *un*usual take the separation of industry from business as a necessary premise. Because of business's dissociation from the material world, Veblen excoriates the immorality of business, and Dreiser celebrates it, yet both positions are enabled by the immaterial basis of business.

48. "Theodore Dreiser Now Turns to High Finance," 194.

3

The Psychology of Desire

Pecuniary Emulation and Invidious Comparison in *Sister Carrie* and *An American Tragedy*

"Without contrast there is no life."

—*Theodore Dreiser*

"No one knows why people want goods. . . . Economists carefully shun the question."

—*Mary Douglas and Baron Isherwood*

The narrator of *The Financier* observes that "The most futile thing in this world is any attempt . . . at exact definition of character" (82). Given the immateriality that Dreiser saw as the basis of business enterprise, it comes as no surprise that he would consider people living within the capitalist order difficult to comprehend. But despite the inscrutability that Dreiser attributes to character, his three most famous fictional creations behave, if not decisively, still in distinctive and memorable ways: Carrie Meeber achieves celebrity on Broadway; George Hurstwood commits suicide in the Bowery; and Clyde Griffiths ends up dead in an electric chair. The fluid personalities as well as the final identities (celebrated actress, pathetic suicide, convicted murderer) of these characters arise from a phenomenon often described as quintessentially American: a fascination with consumer objects so intense that individuality comes to seem, paradoxically, mass-produced.

As Paul Goldberger phrases the familiar lament in a 1997 article for the *New York Times Magazine,* "stores often market themselves as vehicles for creative expression: shop at the Gap and be yourself, buy your furniture at Pottery Barn so that you can discover the real you. But is the real you quite so much like the real me?"[1]

Dreiser's characters provide exemplary illustrations of how that figure much-discredited in recent years as an essentialist and exceptionalist fiction, "the American self," is produced and possibly destroyed in the marketplace.[2] Understanding the fictitiousness of the production of American identity, Dreiser also comprehended its social function. Particularly in *Sister Carrie* (1900) and *An American Tragedy* (1925), he shows how capitalism creates quasi-individual identity (following his working title for the later novel, we might call it the "mirage" of individuality) by endlessly producing desire and eroding identity once desire ceases. Dreiser understood that while desire—for specific consumer goods or for a superior class position—seems to emanate from the individual, it is in fact socially produced.

The most familiar aspect of Veblen's social theory analyzes precisely this method of self-construction. Familiar but misunderstood: Veblen's analysis of the role of consumption in United States life is almost as widely trivialized as it is cited. Ironically, Veblen's panache as a phrasemaker has obscured the richness of his analysis: what he calls "conspicuous consumption" has come to seem self-evident, little more than a status competition against the proverbial Joneses. Although Veblen put consumption studies on the academic map, he has repeatedly been faulted for not seeing how consumers "mak[e] visible and stable the categories of culture" and create "cultural meaning from material objects."[3] To the contrary, he attributes considerable cultural meaning

---

1. Paul Goldberger, "The Sameness of Things," 60.
2. On changing notions of the self in this period, see T. J. Jackson Lears, "From Salvation to Self-Realization," and Jean-Christophe Agnew, "The Consuming Vision of Henry James" (both in the landmark collection, *The Culture of Consumption: Critical Essays in American History,* ed. Richard Wightman Fox and T. J. Jackson Lears); also see T. J. Jackson Lears's *No Place of Grace: Antimodernism and the Transformation of American Culture 1880-1920,* especially 17-18.
3. Mary Douglas and Baron Isherwood, *The World of Goods: Towards an Anthropology of Consumption,* 38; Jackson Lears, *Fables of Abundance: A Cultural History of Advertising in America,* 4. Both of these fine studies err in reducing Veblen's ideas to no more than social status, perhaps because they refer only to *Leisure Class.* In the introduction to *The Sex of Things: Gender and Consumption in Historical Perspective,* Victoria de Grazia praises Veblen as a "deeply committed feminist who regarded female

to consumption, far more than can be reduced to the most familiar Veblenian sound bite. Although much of consumption is indeed conspicuous, Veblen's analysis of it combines psychological complexity (explaining how people see themselves and why they want what they want), sociological insight (relating individual actions to their social order), and theoretical sophistication (incisively rewriting the abstraction of "Economic Man").

The anthropologist Grant McCracken has characterized Veblen as the "founding father" of "person-object relations," a title he should share with Dreiser.[4] They are two of the earliest and most astute chroniclers of the shift from a production-oriented economy to a consumption-driven one. This shift, so familiar in economic history, produced what I will call here the psychology of desire. Once individuals start constructing favorable self-images by purchasing products intended to have a short shelf life, capitalist institutions (firms producing goods, advertising agencies selling goods, mass media distributing the advertising, and so on) grow increasingly powerful and, more significant, increasingly difficult to dislodge.

No wonder that consumer culture has been the subject of protracted (and until recently, almost uniformly negative) discussion. Interestingly, Christopher Lasch characterizes the "propaganda of commodities" as an obstacle to cultural criticism, in that consumption becomes "an alternative to protest or rebellion."[5] Veblen and Dreiser focus less on the damage caused individuals than the divisive effects on society. Noting the uneven distribution of the goods so extensively produced in early-twentieth-century America, Veblen remarks, "The existing system has not made, and does not tend to make, the industrious poor poorer as measured *absolutely* in means of livelihood; but it does tend to make them *relatively* poorer, in their own eyes, as measured in terms of

---

conspicuous consumption as unfree activity" but claims he did not understand "that capitalist society is a semiotic as well as an economic system" (20, 21). Colin Campbell's "Conspicuous Confusion? A Critique of Veblen's Theory of Conspicuous Consumption" concludes, "Veblen's most famous concept is insufficiently clear in its formulation to permit any general agreement on its definition" (45).

4. Grant McCracken, *Culture and Consumption: New Approaches to the Symbolic Character of Consumer Goods and Activities*, 89.

5. Christopher Lasch, *The Culture of Narcissism*, 138. In "Soft Sell: Marketing Rhetoric in Feminist Criticism," Rachel Bowlby describes the new, rehabilitative view of consumption: "Shopping is no longer seen as the despised symptom of patriarchal or capitalist alienation, but rather as part of a newly legitimate . . . 'postmodern' interest in pleasure and fantasy" (382).

comparative economic importance, and, curious as it may seem at first sight, that is what seems to count" (*POS,* 392, emphasis added). Capitalism, that is, makes people feel poorer in comparison with others, even when their absolute standard of living is on the rise. Although this problem is certainly economic, it is also psychological. As Veblen notes, the reason lies in the comparative reckoning of one's self with wealthier others. Thus, since the "aim . . . is not any absolute degree of comfort or of excellence, no advance in the average well-being of the community can end the struggle or lessen the strain." In other words, "A general amelioration cannot quiet the unrest whose source is the craving of everybody to compare favorably with his neighbor" (*POS,* 396–97).[6]

Dreiser also affirms that the relativity of poverty and wealth causes people happiness or misery. He remarks in an 1896 essay that Americans are the "richest people on earth"—yet, as he later says in *Hey, Rub-a-Dub-Dub* (1920), "[i]f the average American has had a little more of food and clothes than the men of some other countries, he has also been confronted by the very irritating spectacle of thousands upon thousands who have so much more than he has or can get" (*SUP,* 49; *HRDD,* 46). Because wealth and poverty are relative matters in the United States, and because there is a premium placed on owning things, identities also become relative to each other.

The relativity of poverty and wealth concerns the cultural critic because of the gap separating the richest cohort from the poorest. Dreiser notes that 350 American families control 95 percent of the wealth; Veblen declares that "something less than ten percent" of Americans "own something more than ninety percent" of all resources (*SUP,* 267; *NOP,* 151). Recent figures show that the inequity persists. A November 1995 *New York Times Magazine* special issue on "The Rich" tracks the total household income of the wealthiest 5 percent of United States families from 1929 to 1993. In 1929, the wealthiest cohort received 30 percent of the total household income, a figure that

---

6. In *The Affluent Society,* John Kenneth Galbraith revisits Veblen's point: "People are poverty-stricken when their income, even if adequate for survival, falls radically behind that of the community. Then they cannot have what the larger community regards as the minimum necessary for decency; and they cannot wholly escape, therefore, the judgment of the larger community that they are indecent. They are degraded for, in the literal sense, they live outside the grades or categories which the community regards as acceptable" (233).

dropped steadily until 1953, when it stopped at less than 16 percent, where it remained until 1983. By 1993, however, the wealthiest 5 percent had increased its receipts to over 20 percent of Americans' total household income. The concentration of total household wealth (as distinguished from yearly income) is even more pronounced. In 1970, the top 1 percent of households owned around 20 percent of the nation's total wealth; by 1989, the share held by the richest 1 percent had grown to around 36 percent. "Welcome back to the past," the *New York Times* writer dryly comments of these figures.[7]

Dreiser's and Veblen's psychology of desire explains the general public's complacency about such patent inequality. Veblen's analysis provides striking insights into how the capitalist order perpetuates itself, which, as in his analysis of business versus industry, furnishes a useful counterpoint to Marx. Rather than positing the inevitable overthrow of capitalism, Veblen assumes its persistence. Whereas Marx describes the bourgeoisie as a sorcerer soon to lose control of the mighty forces he has conjured, Veblen sees the vested interests as bloated parasites that will continue to prey on citizens for the calculable future.[8] A central tenet of Veblen's economic theory is that one host on which business preys is industry. But Veblen also insists that for capitalism to endure, it needs the endorsement of the individuals it exploits; in other words, business preys upon the minds of citizens.

For instance, he describes the point of view of the average United States citizen toward huge conglomerates such as the Standard Oil Company as a "picturesque hallucination that any unearned income which goes to those vested interests whose central office is in New Jersey is paid to himself in some underhand way, while the gains of those vested interests that are domiciled in Canada are obviously a grievous net loss to him" (*VI*, 133). As Veblen astutely points out, the characteristically American veneration for moneymaking follows the same logic as imperialism. American citizens typically admire businessmen because they "like to believe that the personages whom they so admire by force of conventional routine are also of some use . . . that they somehow contribute . . . to the material well-being at large" (*AO*, 116). Similarly, people support imperialism because "by a curious twist of patriotic emotion the loyal citizens are enabled to believe that these

---
7. Andrew Hacker, "Who They Are," 71.
8. For the sorcerer image see Karl Marx, *Manifesto of the Communist Party*, 478.

extra territorial gains of the country's business men will somehow benefit the community at large" (*AO,* 36). Thus when it comes to "enforc[ing] or defend[ing] the businesslike right of particular vested interests to get something for nothing," the individual not only "pays the cost" but "swells with pride" (*VI,* 137).[9]

This emphasis accounts for why Max Lerner describes Veblen's "theory of power" as a "psychological theory of the readiness of the victim for the slaughter," and why T. J. Jackson Lears credits him with anticipating Antonio Gramsci's idea of hegemony. One of Gramsci's most influential concepts is that hegemonic control derives not simply from forcing lower classes to obey the desires of those with power but, in effect, securing their consent to enslave them.[10] Veblen's similar idea of voluntary social control appeared as early as in *The Theory of the Leisure Class* (1899). To use Veblen's term, the average citizen's "hallucination" that capitalism furthers both his personal interest and the nation's collective good helps to ensure business as usual and the perpetuation of a hierarchical social order.[11]

Dreiser articulates a similar, if less developed, notion of hegemony. Echoing Veblen's complaint about self-deluded citizens supporting Standard Oil, Dreiser attributes the American workers' acquiescence "possibly" to their subscription to "the asinine notion in America that every one has an equal opportunity to become a money master, a Morgan or a Rockefeller" (*SUP,* 269). By the cool light of logic it may be easy to see how "asinine" is the view that you or I could ever make

---

9. The common person's happily paying the cost of war is a keynote of Veblen's *An Inquiry into the Nature of Peace and the Terms of Its Perpetuation.*

10. Max Lerner, introduction to *The Portable Veblen,* 27; T. J. Jackson Lears, "Beyond Veblen: Rethinking Consumer Culture in America," 74; Antonio Gramsci, *Prison Notebooks* and *Selections from the Prison Notebooks.* Raymond Williams explains hegemony in *Marxism and Literature,* 108-14. Unlike in Veblen's analysis, Gramsci's concept of hegemony incorporates the possibility of transformation from within the social structure. For a useful distinction between hegemony and coercion, see Geoff Eley, "Nations, Publics, and Political Cultures: Placing Habermas in the Nineteenth Century," 320-23.

11. Veblen's psychological model bears comparison with the Marxist analysis of reification, as developed especially by Georg Lukács. Veblen might agree with Marx's claim that "A commodity is a mysterious thing" but not Marx's assertion that it is also "in the first place, an object outside of us" (Marx quoted in Fredric Jameson, *Marxism and Form,* 295, and in Diggins, *Bard,* 238, n 27). The model of reification, whereby objects in the external world come to seem, in Jameson's words, "frozen" and "static" (*Marxism,* 188) is the opposite of Veblen's model of ever-fluctuating pecuniary emulation and invidious comparison. Diggins astutely distinguishes Veblen's view from the Marxist notion of reification in *Bard,* 101-2.

Rockefeller's millions, but it proves difficult for many United States citizens to disengage themselves emotionally and psychologically from the idea. Thus in *Sister Carrie*, the drummer, Charles Drouet, tells Carrie about the company he works for with "pride in his voice. He felt that it was something to be connected with such a place, and he made her feel that way" (12). And in *An American Tragedy*, a young Clyde Griffiths longs to be "so fortunate as to connect himself with such an institution as" the Green-Davidson Hotel (33). The psychology of desire ensures institutional stability along with endless individual motion.

Veblen's and Dreiser's Psychology: An Overview

Neither Dreiser nor Veblen has received his due as a psychologically astute thinker. As Veblen's analysis of business depressions and stock market fluctuations shows, he understood capitalism to have a psychological as well as an economic foundation. Indeed, Veblen insists, "any theory of culture . . . must have recourse to a psychological analysis, since all culture is substantially a psychological phenomenon" (*ECO*, 17). A 1919 writer for *The Dial* described Veblen as "the Freud of economics" because he emphasizes "psychological forces in the maladjustments of industry and in the economic attitudes of certain classes," and formulates what is, in effect, an "economic libido." The significance of Veblen's foray into psychology is suggested by Gilbert Slater's remark in 1923 that "professed economists, as a rule, appear to regard psychological theory as a Slough of Despond into which it is dangerous to enter, and from which there is no chance of emerging."[12] Veblen enters that Slough of Despond. He insists economists must look at "the living items concerned" rather than assume the abstracted, "normalised scheme[s]" favored by his predecessors (*POS*, 67). His "endeavor . . . to make economics a science" necessitates a more protean theory of human psychology, one that could correspond to the prime "postulate" of "modern" science, "consecutive change" (*POS*, 32).[13] As Veblen

---

12. Vergil Jordan quoted in Dorfman, *Thorstein Veblen*, 421; Gilbert Slater, "The Psychological Basis of Economic Theory," 278. Advertising further suggests the psychological base of business, as Veblen was one of the first to observe. He notes that advertising induces, in particular, fear and shame to encourage sales (*AO*, 310).

13. Veblen's comment about wanting to make economics scientific occurs in a letter to Jacques Loeb, quoted in Dorfman, "New Light," 254. The original letter, dated March 24, 1905, is in the Jacques Loeb Papers, Manuscript Division, Library of Congress.

concisely puts it, his goal is to explain "what men do and how and why they do it" (*POS,* 312).

That same question of motivation preoccupied Dreiser and generations of his critics, especially in assessing Carrie Meeber and Clyde Griffiths. Most readings assume the relationship between psychology and economics to be unidirectional and negative—that the characters are fragmented or destroyed by their materialistic desires. For instance, Dreiser's characters seem "hollow," concludes Sandy Petrey in a landmark essay on *Sister Carrie,* while the eminent Marxist Fredric Jameson finds "commodity lust" characterizing Dreiser himself. James Livingston describes this critical bent as part of "the assumption that there is a necessary contradiction between the development of capitalism and the development of character."[14]

Such critical assessments point also to another sort of trend, for naturalistic characters have traditionally been interpreted as having, in Richard Chase's succinct formulation, "no self." Thus, Chase argues, the typical naturalist character is "[p]sychologically . . . simplified." Thomas Riggio isolates as a prevailing problem in the critical literature on Dreiser just this sort of reasoning; Riggio asks that the judgment of Dreiser's "psychological naivete" be reconsidered and calls for a new look at Dreiser on the grounds of his psychological realism. A promising recent trend, as seen particularly in the work of Philip Fisher, Alan Trachtenberg, and Walter Benn Michaels, focuses less on the commodities Dreiser's characters want and more on why they want them, so now their psychological complexity becomes evident.[15]

---

14. Sandy Petrey, "The Language of Realism, the Language of False Consciousness: A Reading of *Sister Carrie,*" 111; Fredric Jameson, *The Political Unconscious: Narrative as a Socially Symbolic Act,* 159; James Livingston, "*Sister Carrie's* Absent Causes," 234. Related arguments include Stanley Corkin, "Sister Carrie and Industrial Life: Objects and the New American Self"; Lee Clark Mitchell, *Determined Fictions: American Literary Naturalism;* Paul Orlov, "Technique as Theme in *An American Tragedy*"; Bowlby, *Just Looking;* Blanche Gelfant, "What More Can Carrie Want? Naturalistic Ways of Consuming Women." According to Kaplan in *Social Construction,* consumption "compensates for the lack of power" in Dreiser's novels (148).

15. Richard Chase, *The American Novel and Its Tradition,* 199; Thomas P. Riggio, "Carrie's Blues," 24. For another rehabilitative treatment of Dreiserian psychology, see Donald Pizer, "Dreiser and the Naturalistic Drama of Consciousness." My reading comes closest to Alan Trachtenberg's "Who Narrates? Dreiser's Presence in *Sister Carrie,*" which concludes that "*consciousness* is precisely what [*Sister Carrie*] is largely about" (101), and argues "Dreiser's great insight . . . concerns the imitativeness of desire" (108); and Philip Fisher's *Hard Facts,* which describes Dreiser's characters as less "self-made" than "half-made" (13), little more than "anticipatory sel[ves]" (159). Michaels argues in

As Veblen championed an updated psychology to transform sterile economic theory into viable modern science, Dreiser developed a new theory of economic behavior to make modern characters psychologically plausible. Both saw psychology and economics as mutually constituting realms and believed the cultural critic must examine them in tandem. Veblenian psychology, particularly the analysis of the human tendency to compare one's self economically with others (invidious comparison) and the resulting behavior (pecuniary emulation), illuminates the method of self-construction seen in Dreiser's fiction. The novels, in turn, provide rich illustrations of those "living items" Veblen claimed to be analyzing.

The relative, fluctuating nature of the self is a consequence of what Veblen calls invidious comparison. He locates the origins of this modern self-in-flux in the differentiation of individual from collective interests that he associates with the advent of the institution of ownership. His anthropology is inexact, especially in its chronology, but critical to his thinking. The early *Homo sapiens* that Veblen calls peaceable savages did not, he claims, apprehend the idea of private ownership; instead, our peaceful ancestors concerned themselves with the welfare of the group. Even when Veblen peers back to the earliest stages of ownership, what he finds illuminates the behavior of Dreiser's characters. The savage's and very early barbarian's sense of "individuality is conceived to cover . . . a pretty wide fringe of facts and objects," including his shadow, reflection in water, tattoos, footprints, and nail parings that make up a "quasi-personal fringe" (*ECO*, 36–37). The "quasi-personal fringe" expands in Dreiser's novels into a vast net including shirtwaists, hotels, hairstyles, and steak dinners. (In chapter 4 I will take up the Veblenian peaceable savage—a figure he admits derives from "psychology rather than from ethnology" [*TLC*, 21]—at more length.)

Veblen claims that invidious behavior became pronounced during the era of predatory barbarianism, when the "genesis of ownership" caused a "marked accentuation of the self-regarding sentiments" (*IOW*, 160). Thus were the roots of the modern self planted. As "self-interest displaces the common good," "differential advantage of the individual

---

Gold Standard, 41, that Dreiser—unlike Howells—identifies character with desire, and that consequently "the distinction between what one is and what one wants tends to disappear." Veblen's works provide a psychological model contemporary with Dreiser that helps to contextualize such striking insights.

as against his neighbours rather than the undifferentiated advantage of the group" comes to the fore (*IOW,* 160, 161). At this point the reigning "bias" becomes "pecuniary," an orientation Veblen describes as "[i]nvidious, personal, emulative, looking to differential values" (*IOW,* 177). The institution of private ownership leads to "invidious comparison between the possessor of the honorific booty and his less successful neighbours within the group" (*TLC,* 27). Property, no longer shared by a collective body that consumes it, becomes an honorific mark of the self who has—and displays—more booty than his neighbors do. That is because " 'more' means not more than the average share, but more than the share of the person who makes the comparison" (*POS,* 397). Or, as Dreiser puts it more appreciatively, "Differences present life's edge and . . . zest. We enjoy or disdain what we have because of contrast with what we do or do not have" (*D,* 332).

Dreiser also notes that self-image is constructed largely by contrast with images of others. In *Notes on Life* he remarks, "In a world that presents nothing if not contrasts, . . . we can only be aware of ourselves because we are confronted by an endless variety of things and persons which are obviously not us" (195). Dreiser knew well from personal experience how one's sense of self-worth—the apparent value of one's self—could fluctuate continually, depending on the comparative wealth or poverty of one's neighbors. Always "sharply aware of what it meant to have less as opposed to more," Dreiser says in a strangely mixed metaphor that he does not "believe any human being ever watched the pleasures of others with a hungrier eye" (*D,* 52, 398). An early memory recorded in *Dawn* (1931) seems to have been formative for Dreiser: "[o]r again, I am passing the residence of a rich man. . . . My face is pressed between two iron pickets of the fence, to see. What could so great an animal be?" (18). The more one looks, the more one sees inequality; thus there will always be—to borrow Veblen's phrase— "progressive rerating" of the self (*TBE,* 234).

The related concept of pecuniary emulation is easily trivialized. As Veblen warns, "The fear of losing prestige has often been confused as personal vanity; a mistake due to incomplete analysis" (*AO,* 310 n17). To comprehend the force of the idea of pecuniary emulation, and to see its usefulness for explaining the behavior of Dreiser's characters (and of many real-life counterparts), we must move beyond using the idea of consumption as a terminal point in the analysis of behavior.

One of Veblen's most frequently overlooked points is that "the motive that lies at the root of ownership is emulation"—not, as it turns out, consumption (*TLC*, 25). In fact, Veblen argues, emulation "is probably the strongest . . . of the economic motives proper," second only to self-preservation, and one of the most "pervading trait[s]" of human psychology (*TLC*, 110, 109). With the evolution of *Homo sapiens*, the "opportunity and incentive" to emulate increases in "scope and urgency," a process only accelerated with the growth of capitalism (*TLC*, 16). Veblen believes "economic emulation . . . pervades the structure of modern society more thoroughly perhaps than any other equally powerful moral factor" (*POS*, 398).

Sister Carrie and *An American Tragedy* illustrate the role emulation plays in motivation: Carrie Meeber, George Hurstwood, and Clyde Griffiths are motivated less to consume desirable goods, as most critics have assumed, than to emulate desirable people. In fact, Dreiser claims in *Sister Carrie* that "the innate trend of the mind. . . . [is] to emulate the more expensively dressed" (49).[16] What is "innate" to the mind of modern man and woman, according to Dreiser, is first to compare one's self (and one's objects) to others, then to improve the self by emulation. *Sister Carrie* provides a microeconomic view of the psychology of desire. Focusing on Carrie's re-creating of herself and enhancing of her personal value by assuming ever more elusive desires, Dreiser constructs a bildungsroman of invidious comparison. Hurstwood illustrates the complementary process: the diminishment of personal value due to defective comparative and emulative strategies. *An American Tragedy*, providing a more macroeconomic view of pecuniary emulation and invidious comparison, offers a more thoroughly institutional perspective. Whereas the earlier novel focuses on individual intent (the acquisition of goods), the latter emphasizes social function (stabilizing capitalism). Perhaps for that reason, *Carrie* portrays the psychology of desire in a more attractive, even seductive, light, while *Tragedy* demonstrates Dreiser's more mature outlook as a cultural critic.

16. Choosing between editions of Dreiser's writing—the works as originally published, or those recently published by the University of Pennsylvania Press—is particularly challenging in the case of *Sister Carrie*, not only because it was the first and most highly publicized of the "restored" texts, but also because the differences are so pointed. I cite the older version more often simply because the shorter length makes it likely to remain more familiar.

The Microeconomic View: Invidious Comparison in *Sister Carrie*

Writing in 1996, Abigail Solomon-Godeau notes, "One of the most conspicuous features of commodity culture is its sexualization of the commodity, its eroticization of objects, which in turn inflects, if not determines, the psychic structures of consumer desire."[17] Dreiser demonstrated that idea nearly one hundred years earlier in his portrayal of Carrie Meeber's love affair with money and consumer goods. The famous passage where Charles Drouet pulls out "soft, green, handsome ten-dollar bills" to seduce Carrie and the scene in which she first feels department store trinkets "touc[h] her with individual desire" suggest that Carrie's eros has been sublimated to acquisition (*SC*, 62, 26). Yet Dreiser maintains that his heroine was "too full of wonder and desire to be greedy" (*SC*, 120). In the Pennsylvania edition, Dreiser begins the chapter following Hurstwood's visit to Drouet and Carrie's flat by cautioning readers against simplistic readings of her psychology: "In considering Carrie's mental state," the narrator warns, "[t]he things that appeal to desire are not always visible objects. Let us not confuse this with selfishness. It is more virtuous than that" (Penn *SC*, 97). Carrie's wondering and desiring relationship to objects is not a matter of greedy consumption but of hungry self-creation. Dreiser's analysis of Carrie's bent toward emulation, which is both more subtle and more sustained than his treatment of her consumption, follows the prime tenets of Veblen's psychological model: observation, contrast, and consecutive change.

Dreiser revises the traditional terms of the bildungsroman so that Carrie's growth occurs not in "knowledge" but in "desire" (*SC*, 112). This bildungsroman of invidious comparison depicts consecutive transformations of the self, as Carrie evolves toward the acquisition of more elegant objects and compares herself to more glamorous people. Immediately in the first chapter of the novel, Drouet's clothes stimulate the "innate trend" of Carrie's emulating mind and "[s]he became conscious of an inequality" (*SC*, 49, 11). The moment marks an epiphany for Carrie, as the contrast between her personal effects and Drouet's

---

17. Abigail Solomon-Godeau, "The Other Side of Venus: The Visual Economy of Feminine Display," 113. The collection in which this essay appears—aptly titled *The Sex of Things*, ed. de Grazia—contains many arguments that intersect with mine.

teaches her the relativity of wealth—which she registers as her personal "inequality." Self and possessions can't be separated, for Carrie is struck by "[h]er insignificance in the presence of so much magnificence" (Penn *SC*, 7). Quickly tallying and comparing, Carrie's self recognizes its shabbiness. In Chicago, the department store instructs Carrie further, as tantalizing objects meet with well-turned-out women, and Carrie submits herself to the painful comparison: "Their clothes were neat, in many instances fine, and wherever she encountered the eye of one it was only to recognize in it a keen analysis of her own position—her individual shortcomings of dress and that shadow of manner which she thought must hang about her and make clear to all who and what she was. A flame of envy lighted in her heart" (*SC*, 27). This moment exemplifies invidious comparison and pecuniary emulation: Carrie compares herself and her objects to others and rapidly reconstructs a new self in her mind. Resolved to enhance her personal value, Carrie sets out to transform "who and what she was."[18]

"[C]ontrast," the narrator remarks in the Pennsylvania edition, "was the proper lever to move [Carrie's] mind" (Penn *SC*, 317). Her development through the novel illustrates what Veblen considers "the normal effect" on an individual living in a pecuniary culture: "[a] prompt response to . . . stimulus" and continually shifting desires (*TLC*, 103). Disputing the hedonistic psychological model favored by orthodox economists and derived from Jeremy Bentham (better known today as the designer of the panopticon discussed by Foucault in *Discipline and Punish*), Veblen argues that people are not "passive and substantially inert." In a 1912 book review, he criticizes G. Tarde's *Psychologie Économique* for emphasizing the "mechanical schematization of phenomena" at the cost of explaining "the springs of human conduct" (*ERR*, 497). His attack on orthodox economic theory includes a sustained critique of its static view of psychology, which Veblen describes as constituting a "theory of valuation with the element of valuation"—that is, the active human agent—"left out" (*POS*, 144). Veblen mocks in particular "[t]he hedonistic conception of man [as] that of a lightening calculator of pleasures and pains, who oscillates like a homogeneous globule of desire of happiness under the impulse

---

18. Carrie's self-construction also resembles what Jean-Christophe Agnew, in "A House of Fiction: Domestic Interiors and the Commodity Aesthetic," describes as "a point of view that celebrates those moments when the very boundaries between the self and the commodity world collapse in the act of purchase" (135).

of stimuli that shift him about the area, but leave him intact" (*POS*, 73). The model of the "isolated," "stable," and normalized Economic Man provided Veblen with another indication that orthodox economists were "helplessly behind the times" (*POS*, 73, 56).[19] Not surprisingly, "economic conduct still continues to be somewhat mysterious to the economists" (*POS*, 178).

Veblen follows instead "[t]he later psychology" in seeing "the characteristic of man [as] to do something, not simply to suffer pleasures and pains" (*POS*, 74).[20] The individual, argues Veblen, does not resemble the hedonist's "self-contained globule of desire" "le[ft] . . . intact" by stimuli; she is "an agent, not an absorbent" (*POS*, 73-74, *ECO*, 81).[21] Generally a fierce opponent of teleological reasoning, Veblen locates in the human agent's "unfolding impulsive activity" the indication of a "'teleological' activity" of sorts: the individual "seek[s] in every act the accomplishment of some concrete, objective, impersonal end"

---

19. In *Thorstein Veblen*, Riesman argues that Veblen's rejection of Benthamite psychology inaugurates "a new direction for economic thought: since man was an agent, he must be considered as a whole, and by the same token all human activity, as part of the context, must be studied by economists if they are to understand narrowly 'economic' behavior" (151). But according to Lears in "Beyond Veblen," while Veblen goes beyond the "simple-minded utilitarianism of orthodox economics," his psychology remains "narrow": "he still reduced complex social rituals to one-dimensional examples of 'pecuniary emulation'" (75).

In *By the Law of Nature*, Horwitz examines the "emulative self," arguing that Veblen provides "an American pragmatist alternative" to the "absolutist premises of classical organicism" (137). Also see Alan W. Dyer, "Semiotics, Economic Development, and the Deconstruction of the Economic Man," and David B. Hamilton, "Institutional Economics and Consumption," especially 121.

20. Veblen's habit of rarely citing sources makes it difficult to trace his influences. Many scholars see the instinct psychology practiced by William James, William McDougall, and Jacques Loeb (the latter a direct influence on Dreiser as well) as shaping Veblen's psychology. On how the "new psychology" differs from the old, see Daniel J. Wilson, *Science, Community, and the Transformation of American Philosophy, 1860-1930*, chapter 5.

William James provides a particularly important reference point. A premise of his *Principles of Psychology* (1890) is that *Homo sapiens* is "*the* imitative animal" (2:408). James reasons that humans have "an innate propensity to get ourselves noticed" and finds the line "between what a man calls *me* and what he simply calls *mine* . . . difficult to draw" (1:293, 291). Thus a dramatic "loss of possessions" causes a "sense of the shrinkage of our personality" (1:293). But James finally posits a "*hierarchical scale, with the bodily Self at the bottom, the spiritual Self at the top, and the extracorporeal material selves and the various social selves between*" (1:313). Veblen rejects the hierarchy, along with the Jamesean "nuclear" self, that "truest, strongest, deepest self" (1:302, 310).

21. For a recent assault on a related construct still dominant in economic theory, see Geoff Hodgson, "Behind Methodological Individualism."

(*TLC*, 15). Veblen's emphasis on human agency grants the consumer more power of selection, and thus more of a role in the production of cultural meanings, than his orthodox competitors allow, or than most of his critics have realized. Although commentators systematically overlook this point, Veblen actually gives back to the consumer some discretionary power.[22]

Veblen does not, however, believe that the ends people seek are necessarily rational. As much as Veblen complains of hedonistic psychology's assumption of static agents, he rails against its utilitarian view that Economic Man rationally and accurately weighs his options. Arguing that "modern science has to do with the facts as they come to hand, not with putative phenomena warily led out from a primordial metaphysical postulate, such as the 'hedonic principle,'" Veblen contests the notion that people rationally seek pleasure and avoid pain (*ECO*, 150; *POS*, 234-35). To the contrary, he points out, invidious comparison causes people to seek out psychological anguish. While "the contemplation of a wealthy neighbor's pecuniary superiority yields painful rather than pleasurable sensations," the observer will hold this wealthy neighbor in higher esteem than "another neighbor who differs from the former only in being less enviable in respect of wealth" (*POS*, 246-47).[23] That is because the observer hopes, much like the patrons of Hurstwood's Chicago saloon, "to rub off gilt from those who have it in abundance" (*SC*, 242). Bringing the pain of invidious comparison upon oneself makes no more sense than looking up to one's neighbors simply because of their wealth, but the reflex is familiar. Dreiser himself once confessed, "I think most of us like to be upstaged at one time or another by some one" (*BM*, 182).

---

22. Veblen explains the individual's teleological behavior: "The discipline of life in a modern community, particularly the industrial life, strongly reënforced by the modern sciences, has divested our knowledge of non-human phenomena of that fullness of self-directing life that was once imputed to them, and has reduced this knowledge to terms of opaque causal sequence. It has thereby narrowed the range of discretionary, teleological action to the human agent alone" (*POS*, 179). Thus, Veblen's attack on animistic thought (a good example is Adam Smith's "invisible hand") commits him to giving discretion back to humans.

23. Contrast Georg Simmel, who in his 1904 essay on "Fashion," says of the poor person who glimpses the feast of a rich neighbor, "The moment we envy an object or person, we are no longer absolutely excluded from it; some relation or other has been established—between both the same psychic content now exists. . . . The quiet personal usurpation of the envied property contains a kind of antidote, which occasionally counteracts the evil effects of this feeling of envy" (14).

"The heart understands," the narrator of *Sister Carrie* remarks, "when it is confronted with contrasts" (304). Contrast drives both Carrie and the plot of the novel forward, contrast that exists within the mind of the individual experiencing it. As Veblen explains, further emphasizing the role of individual agency, "The physical properties of the materials accessible to man are constants: it is the human agent that changes,—his insight and his appreciation of what these things can be used for is what develops." Further, "the constitution of the organism, as well as its attitude at the moment . . . , in great part decides what will serve as a stimulus" (*POS*, 71, 155). Carrie's reaction to her sister Minnie's flat after her first visit to Fitzgerald and Moy's (Hannah and Hogg's in the Pennsylvania edition) illustrates Veblen's point: "Carrie felt a new phase of its atmosphere. The fact that it was unchanged, while her feelings were different, increased her knowledge of its character" (51). The stimulus that prompts the mental transformation, whether it is the atmosphere of Minnie's flat or the shine of Drouet's shoes, does not change; Carrie herself changes. Minnie's comment that "Carrie doesn't seem to like her place very well" (Penn *SC*, 50) is more perceptive than she may realize: Carrie likes neither her place at the shoe factory nor her social place living with the Hansons. Because Carrie's attitudes change so rapidly, she is the darling of invidious comparison.[24]

Carrie proves a genius at this method of self-construction. Other women whom Drouet finds attractive prompt her to emulate rather than to envy them: "If that was so fine, she must look at it more closely. Instinctively, she felt a desire to imitate it" (*SC*, 100). John Berger observes, "Women watch themselves being looked at," but Carrie is

---

24. Occasionally Dreiser endorses the older, hedonistic psychology: "All that the individual imagines in contemplating a dazzling, alluring, or disturbing spectacle is created more by the spectacle than the mind observing it. . . . We are, after all, more passive than active, more mirrors than engines" (Penn *SC*, 78). On Dreiser endorsing a hedonistic or Spencerian psychological model, see Lehan, "The City," and Noble, "Dreiser and Veblen." Consider also H. L. Mencken's comment in "Dreiser in 840 Pages" that while Dreiser thinks he is a Freudian, "he is actually a behaviorist of the most advanced wing" (799-800).

Dreiser's reputation for creating passive characters might seem to contradict my argument. Hurstwood, while indeed passive, manifests that passivity in an inability to construct a self through invidious emulation. Carrie and Clyde are not really passive—for they do go after what they want, whether nice clothes or wealthy lovers—even though their desires can seem irrational, tawdry, or silly to observers who place themselves outside of (and, typically, above) the fictional world (see also Warwick Wadlington, "Pathos and Dreiser," 222).

more apt to look at other women being observed.[25] She extends her comparisons from beautiful women to attractive settings, as when, after a carriage ride with Mrs. Hale through Chicago's fashionable section, "she came to her own rooms, [and] Carrie saw their comparative insignificance. . . . She was not contrasting it now with what she had had, but what she had so recently seen" (*SC*, 113). The qualification emphasizes her expertise in invidious comparison; unlike Hurstwood, Carrie looks forward in time and upward in status to construct herself. She changes so rapidly that when Hurstwood calls on Carrie while Drouet is out of town, he "found a young woman who was much more than the Carrie to whom Drouet had first spoken" (*SC*, 103). Being so "naturally imitative" (*SC*, 103), Carrie can keep constructing more and more successful selves.

The method of self-fashioning described by Dreiser and Veblen depends as much upon being observed by others as on observing them; what matters is display. Recent work in film theory and cultural studies debates the idea of a "male gaze" that allegedly fixes women as objects and denies them subject positions, but Dreiser and Veblen's treatment of visibility suggests there may be a capitalist gaze.[26] Carrie, in fact, is introduced as having a distinctive gaze ("[a] clever companion . . . would have warned her never to look a man in the eyes so steadily" [*SC*, 12]), while "Drouet was all eyes" (Penn *SC*, 101). This capitalist gaze does not so much render people as objects as it forces people to see themselves in relation to objects, and in fact see themselves being seen with objects. Veblen's memorable phrasing draws attention to exactly what he has been accused of ignoring—that consumer items form a semiotic system clearly intelligible to people in a given culture: "[i]n order to impress these transient observers, and to retain one's self-complacency under their observation, the signature of one's pecuniary

---

25. John Berger, *Ways of Seeing*, 47. Notably, when Carrie does feel envy, as in her first trip to the department store, it is not toward the elegant ladies but their clothes. Carrie knows "she now compared poorly" (*SC*, 27).

26. In *Ways of Seeing*, Berger succinctly describes the male gaze: "In the average European oil painting of the nude the principal protagonist is never painted. He is the spectator . . . and he is presumed to be a man" (54). In consequence, women are "split into two. A woman must continually watch herself" (46) and "survey, like men, their own femininity" (63). On the gaze, see Laura Mulvey, *Visual and Other Pleasures;* Mary Ann Doane, *The Desire to Desire: The Woman's Film of the 1940s;* and Teresa de Lauretis, *Alice Doesn't: Feminism, Semiotics, Cinema*. And see Bowlby's *Just Looking* for further discussion of looking and display in Dreiser's novels.

strength should be written in characters which he who runs may read" (*TLC*, 87). The reason is that "esteem is awarded only on evidence" (*TLC*, 36). Dreiser makes the same point in the Pennsylvania *Sister Carrie*: "how dismal is progress without publicity. . . . [I]ndividuals love more to bask in the sunshine of popularity than they do to improve in some obscure intellectual shade. Merit is no object, conspicuity all" (173). Once the importance of visibility is appreciated, it becomes clear that it is not wealth per se that impresses Dreiser's characters but "evidences of wealth" (*SC*, 36).[27]

As Dreiser describes the perspective of a character in another novel, "appearances were worth something" and thus, "[c]lothes were the main persuasion" (*JG*, 11, 12). Clothes and other personal furnishings are so persuasive that Drouet's talk to Carrie; when he serves her first dinner in a Chicago restaurant, "his rings almost spoke" (*SC*, 60). Hurstwood will similarly advance his suit through this powerful language of personal property; Carrie listens, not to "his words, [but to] the voices of the things which he represented" (*SC*, 115).[28] As the narrator remarks early on, "A woman should some day write the complete philosophy of clothes" (*SC*, 10). In fact, many feminist theorists have been doing just that of late. Leslie W. Rabine argues that "fashion does not merely express this [feminine] self, but, as a powerful symbolic system, is a major force in producing it." Kaja Silverman goes so far as to affirm "that clothing is a necessary condition of subjectivity."[29] The sartorial

27. As the Pennsylvania edition makes clear, Carrie initially shuns conspicuousness in Chicago. Seeking work, she tries "to avoid conspicuity"; when "seeing herself observed, [she] retreated" because she feels more comfortable "unobserved" (Penn *SC*, 18, 19). On the job, she cultivates "inconspicuousness" (Penn *SC*, 40). But even her initial avoidance of conspicuously displaying herself—a behavior that quickly changes—emphasizes Carrie's aptitude for pecuniary emulation and invidious comparison, as she has not yet placed herself to compare favorably.

28. As is often the case, Dreiser draws directly from his own experience. In *Dawn*, he records a job he held with "an immense wholesale hardware company" in Chicago—he describes it as "truly a school of economics and, in its way, sociology." Young Dreiser was amazed by the piles of goods: "ironware, tinware, rivets, nails, brooms, mops, can-openers, glass jars, razors, shaving mugs and brushes, spoons, ladles, coal scuttles, fire sets, fire buckets, kitchenware." Yet more remarkable than the sheer accumulation of goods is their penchant for talking to Dreiser: " 'You need me! You need me! You need me!' " they cry out (*Dawn*, 335, 338-39).

29. Leslie W. Rabine, "A Woman's Two Bodies: Fashion Magazines, Consumerism, and Feminism," 59-60; Kaja Silverman, "Fragments of a Fashionable Discourse," 191. Many of the essays collected with these in *On Fashion*, ed. Shari Benstock and Suzanne Ferris, likewise affirm connections among fashion, identity (especially female identity), and consumerism.

aspirations of Carrie and other Dreiser characters testify to the cultural and personal levels of meaning that recent feminist theorists attribute to fashion.

Dreiser contributes an important chapter to the philosophy of clothes when he remarks on Carrie's "need of clothes—to say nothing of her desire for ornaments" (361). This distinction follows Veblen's analysis of the utilitarian function of clothing versus the ceremonial role of dress. Clothing keeps the body warm and protected; dress ministers largely to invidious self-esteem. Most people, notes Veblen, will "go ill clad in order to appear well dressed" because "the chief element of value in many articles of apparel is not their efficiency for protecting the body, but for protecting the wearer's respectability" (*TLC*, 168; *POS*, 395). In a brilliant essay developing "The Economic Theory of Woman's Dress" that he later incorporated into *Leisure Class*, Veblen locates three "cardinal principles"—novelty, cumbersomeness, and costliness—that argue against the instrumental function of clothes (*ECO*, 74-75). First, the novelty of a fashion, giving evidence of "a constant supersession of the wasteful garment . . . by a new one" signifies "conspicuous waste" (*ECO*, 72). Second, by incapacitating the wearer from useful effort, fashion enhances her prestige by testifying to her conspicuous leisure.[30] And finally, of course, fashions must be expensive in order to demonstrate wealth. Carrie apprehends Veblen's distinction between clothing and dress even before she moves in with Drouet: "She could possibly have conquered the fear of hunger and gone back" to the factory—"but spoil her appearance—be old-clothed and poor-appearing—never" (Penn *SC*, 99). Carrie knows she "need[s] more and better clothes to compare" (*SC*, 288).

Probably no novelist better documents the American obsession with dress to enhance personal value than Dreiser, and it is not only his fictional characters who understand the uses of dress to construct reputable selves. Dreiser describes in moving detail in *An Amateur Laborer* (published posthumously in 1983) his nervous breakdown

---

30. I use the feminine third person singular in conformity with Veblen's analysis. In the case of women's clothing, as with servants', Veblen distinguishes between the wearer (the wife) and the owner (the husband); "But while they need not be united in the same person, they must be organic members of the same economic unit" (*ECO*, 67).

Veblen's emphasis on dress as signifying reputability inverts the pattern identified by Mariana Valverde in "The Love of Finery: Fashion and the Fallen Woman in Nineteenth-Century Social Discourse." Valverde's argument is interesting, although she imposes on Dreiser a conventional moralism toward female sexuality that he does not exhibit.

following the "suppression" of *Sister Carrie* and the death of his father. Down to his last dollar, depressed, and exhausted, Dreiser refuses to pawn his watch or hatbox, because "I would be destroying my chances . . . if I began to part with the things that made up my personal appearance" (*AL,* 28). The sympathy of one who has "been there" permeates Dreiser's treatment of Carrie's love of clothes.

To prevent us from seeing Carrie as, say, just an insect particularly well adapted to its tree branch, Dreiser lovingly declares that she has "more imagination" than the working girls at the Chicago shoe factory (*SC,* 54). Her superior imagination manifests itself in the continual reconstruction of prettier, more desirable Carries. Imitation becomes, appropriately enough, her art form. Indeed, Drouet assumes she can perform on stage because he has seen her "giving imitations" at home (*SC,* 149). Carrie's "innate taste for imitation" (*SC,* 150) draws her to acting. Further linking Carrie's artistic ambitions with her skill at invidious comparison and pecuniary emulation, Dreiser explains that the "[d]ramatic art" "inspires thoughts of emulation in . . . observers" (Penn *SC,* 158). Carrie's success on the stage, first in the amateur theatre in Chicago and later as a New York professional, and her striving at the end of the novel toward "such happiness as [she] may never feel" (*SC,* 465) have traditionally been read as indications of some empty or failed self. But the bildungsroman of invidious comparison cannot end in the transcendent wisdom or stable marriage of its heroine; Dreiser culminates his story, appropriately, with the apotheosis of desire.

Interpretations of *Sister Carrie* as showing Dreiser's critique of consumption frequently pivot on the analysis of Bob Ames. This character indirectly sheds light on Dreiser's cultural criticism. Many readers feel that Ames, like Hurstwood or the Hansons, encourages Carrie *not* to desire. Ames would not, he says, care to be rich; while eating at Sherry's Restaurant he claims that wealthy people confuse their ability to pay with the actual worth of goods (*SC,* 301, 299). Ames's productivist ethic suits his profession as an electrical engineer, but the similarity to Veblen's romanticized engineers in *The Engineers and the Price System* (1921) ends there. Although Carrie evolves from favoring representatives of two professions that Veblen rates wasteful and futile (salesman and saloon manager)[31] to preferring an engineer, Bob Ames

---

31. Saloons are "typical of a class of investments which derive profits from capital goods devoted to uses that are altogether dubious, with a large presumption of net

does not provide a viable alternative. As a cultural critic, Ames is a failure. Only temporarily does he "tak[e] away some of the bitterness of the contrast between this life [represented by the opulence at Sherry's] and [Carrie's] life" (*SC,* 301-2). Far from restraining Carrie's invidiousness, Ames becomes, ironically enough, her new "ideal to contrast men by" (*SC,* 304).

Ames cannot escape the logic of invidious comparison that seems increasingly to seduce Dreiser in the second half of the novel. Ames's advice reinforces Carrie's pecuniary, emulative, and comparative methods of reckoning. He recommends that she make herself "valuable to others" as a way to "make [her] powers endure" (*SC,* 448): seek happiness, he counsels, by relying on others to validate your worth. But this is the same process Carrie has always used to "increas[e]" her personal "value" (*SC,* 143). The "burden of duty" (*SC,* 448) that Ames wants Carrie to bear is the burden of encouraging others to desire. In fact, Ames tells Carrie that her face is "representative of all desire" (*SC,* 448). He encourages Carrie to stimulate in her devotees what has been identified as the "unique 'spirit' of modern consumerism": the "obligation and duty" of "dissatisfaction and desire." "At root," says Colin Campbell, modern consumerism is "based upon a strong sense of duty, an obligation to engage in 'want satisfaction' as an end in itself." Each time she acts on stage, Carrie not only reenacts herself through making believe but also incites others to desire. Her roles elicit desire but, because they are roles, can never satiate it. The only engineering we see Ames do is to lubricate the wheels of the economy of desire.[32]

Dreiser ends the bildungsroman of invidious comparison with Carrie successful, yet still yearning, to illustrate the unquenchable and

---

detriment" (*POS,* 357). Veblen lets loose on salesmanship numerous times; see, for instance, *Engineers,* 113, where he reckons it doubles the price of goods and services, and *Absentee Ownership,* 284-325.

32. Colin Campbell, "Romanticism and the Consumer Ethic: Intimations of a Weber-Style Thesis," 282, 284.

Desire is a perennial topic in Dreiser criticism, as seen in Randolph Bourne, "The Art of Theodore Dreiser"; Warren, *Homage;* Fred G. See, *Desire and the Sign: Nineteenth-Century American Fiction;* Hussman, *Dreiser and His Fiction;* Hochman, "A Portrait of the Artist"; Livingston, "Absent Causes"; Hurwitz, *By the Law of Nature;* and Leonard Cassuto, "Dreiser's Ideal of Balance."

Bersani's *A Future for Astyanax* set an important critical trend by identifying realist novels with a model of character that attempts to contain desire. Dreiser's characters tend to contradict Bersani's provocative argument—for instance, that "Realistic fiction admits heroes of desire in order to submit them to ceremonies of expulsion" (67).

strangely immaterial quality of desire. Carrie enjoys "that middle state" when "possessed of the means" to increase her store of goods, "lured by desire, and yet deterred" (*SC,* 67). Since the end sought is not consumption but comparison and emulation, satiation is impossible. Indeed, says Veblen, "emulation in expenditure stands ever ready to absorb any margin of income. . . . [I]n the long run, a large expenditure comes no nearer satisfying the desire than a smaller one" (*POS,* 394-95). Or, as Dreiser generalizes in *Dawn,* "Here was no land or day to be satisfied with well enough" (293). Consumption must end, but desire cannot. Stable desire, for both Veblen and Dreiser, is an oxymoron.[33]

Carrie, her "craving for pleasure" providing the "one stay of her nature," represents the paradox of desire (*SC,* 34). "[T]he desireful Carrie,—unsatisfied" illustrates Veblen's assertion that "the standard . . . which commonly guides our efforts is not the average, ordinary expenditure already achieved; it is an ideal of consumption that lies just beyond our reach" (Penn *SC,* 487; *TLC,* 103). Little wonder that to Carrie, people who "seemed satisfied with their lot" appear "'common'" (Penn *SC,* 53). As Veblen explains, one's standard of living "is of a very elastic nature, capable of an indefinite extension"; thus, "want . . . is indefinitely expansible, after the manner commonly imputed . . . to higher or spiritual wants" (*POS,* 394; *TLC,* 111). The psychology of desire also accounts for what Jackson Lears describes as the "dematerializing of desire" propagated by twentieth-century advertising, which entices the individual to seek not "satisfactions of actual possession . . . [but the] excitement of anticipated purchase."[34] One reason desire becomes dematerialized is that the consumption of goods matters less than the construction of an image. The desires that seem eminently microeconomic matters are, equally, psychological ones.

Hurstwood and the Tragedy of Noninvidiousness

"We talk of stable things, but nothing is stable," Dreiser remarks in *Dawn.* "If you let things alone, they change" (371). As the career

---

33. The idea that desire is limitless has a precedent in Darwin, who says in *The Descent of Man and Selection in Relation to Sex* that "the wish for another man's property is perhaps as persistent a desire as any that can be named; but even in this case the satisfaction of actual possession is generally a weaker feeling than the desire" (90).

34. Lears, *Fables,* 49, 51.

of George Hurstwood reminds us, change need not mean progress. Ever since H. L. Mencken declared that the Hurstwood story "broke the back" of *Sister Carrie,* numerous critics have seen the two plots as conflicting.[35] Reading the novel as providing two illustrations of Veblen's psychological model (with Carrie a success at invidious comparison and Hurstwood a failure at it) provides one way of resolving the supposed conflict. Initially, invidious comparison works to Hurstwood's advantage. The middle-aged suitor appears to Carrie "more clever than Drouet in a hundred ways," quicker with money when the three play euchre for dimes, and obviously, she concludes, "the superior man" (*SC,* 94, 96–97, 108). Drouet, now "dull in comparison" with Hurstwood (*SC,* 108), has no more changed than Minnie's flat did earlier, but Carrie is ready to pursue a better object of desire.

Hurstwood's interest in the starlet is also stimulated by invidious comparison. The Pennsylvania edition makes clear that the affair with Carrie is not his first; other young women have made Mrs. Hurstwood appear "deficient by contrast" (Penn *SC,* 85). Why Hurstwood's erotic longings lead to his destruction becomes clear when interpreted in light of the Veblenian psychology of desire. What Hurstwood initially seeks from Carrie is someone to "waste a little affection on me" (*SC,* 124). Although Veblen avoids discussing sexual or romantic impulses, the concept of waste recurs throughout his thinking. He consistently condemns waste on ethical grounds, while arguing that ceremonial waste confers honor on those who practice it. For instance, Veblen notes ironically, "The possession of an heirloom is to be commended because it argues the practice of waste through more than one generation" (*ECO,* 75). Displaying a propensity for waste, therefore, confers honor. In Hurstwood's case, getting sexual and affectional "waste" from Carrie would likewise help him to construct a more conspicuously reputable self.

But Hurstwood's sexual passion for Carrie is also quite literally self-destructive. He wrongly assumes an affair will follow the hedonistic calculus, thinking a fling with Carrie "represented only so much added pleasure. He would enjoy his new gift over and above his ordinary allowance of pleasure. He would be happy with her and his own affairs would go on as they had—undisturbed" (Penn *SC,* 133). What Hurstwood fails to realize is that one cannot introduce a change and remain

---

35. H. L. Mencken, "A Novel of the First Rank," 742.

"undisturbed." His mistaken assumption that he can indulge his passion for Carrie without disturbing his life foreshadows his downward spiral once so many other factors change in New York. Hurstwood's problem is that he would be content to be that self-contained globule of desire that both Dreiser and Veblen felt had been superannuated.[36]

Hurstwood's immediate and radical status change in New York indicates that his manhood and selfhood are subject to sudden change for the worse. The narrator remarks that "[w]hatever a man like Hurstwood could be in Chicago, it is very evident that he would be but an inconspicuous drop in an ocean like New York. . . . A common fish must needs disappear from view. In other words, Hurstwood was nothing" (273). This concise description illustrates Hurstwood's regression from man to fish to the final term: nothing. Comparatively speaking, Hurstwood in New York is no longer a self. He admits as much when he applies for a scab job with the streetcar company. Asked if he's a motorman, Hurstwood ominously answers, "No, I'm not anything" (Penn SC, 413). Hurstwood's fall, in the words of Ellen Moers, from "sham celebrity to fatal anonymity" occurs due to his defective psychology of desire.[37]

One of the most poignant aspects of the Hurstwood saga is his maintaining a sense of personal superiority even when the facts argue otherwise. His superiority complex emerges, for instance, when he takes the scab job but feels "a little superior to these two [scab workers]—a little better off" (SC, 381). Hurstwood's snobbery continues when, appealing to a hotel porter for scrub work and "recognizing even in his plight the man's inferiority to him," he tells of his old managerial job and "the figure of Hurstwood was rather surprising in contrast to the fact" (SC, 426, 427). That his static sense of self anchors Hurstwood to his past causes him considerable trouble. Since "people took him to be better off than he was, . . . it retarded his search" for a job in New York (SC, 319). Hurstwood's playacting at gentility, which the narrator describes as his adopting a "feeble imitation" of his past success, dooms him to failure (SC, 320). Hurstwood's reliance on times past to constitute a sense of self culminates in the move to the fifteen-cent lodging house, where "his preference was to close his eyes and dream of other days. . . . As the present became darker, the past grew brighter, and all that concerned

---

36. See also Hussman, *Dreiser*: "The unmistakable symptom of [Hurstwood's] disease is the waning of desire" (24).

37. Moers, *Two Dreisers*, 103.

it stood in relief" (*SC,* 424-25). Virtually since arriving in New York, Hurstwood has been indulging in invidious "comparisons between his old state and his new [which] showed a balance for the worse" (*SC,* 305). Unfortunately for Hurstwood, "contrasting his present state with his former . . . became a natural method of mentation with him—to think of doing a thing now, and then quickly remembering how he did it formerly" (Penn *SC,* 311). As the narrator foretells Hurstwood's fate early in the New York section, "This was not the easy Hurstwood of Chicago. . . . The change was too obvious to escape detection" (Penn *SC,* 311).

Places, particularly cities, exercise powerful effects on Dreiser's characters. He depicts Chicago itself as a site of invidious comparison and not only because its department stores encourage consumption and incite desire. Chicago, a "giant magnet," looks toward the future just as Carrie does: "Its population was not so much thriving upon established commerce as upon the industries which prepared for the arrival of others" (Penn *SC,* 16). Chicago also magnifies contrast: "The entire metropolitan centre possessed a high and mighty air calculated to overawe and abash the common applicant, and to make the gulf between poverty and success seem both wide and deep" (*SC,* 17).[38]

Although Chicago enhances Carrie's invidious self-construction, New York is not to blame for Hurstwood's decline into self-annihilation. His problem, as manifested in various ways well before he moves east, is that he is no good at invidious comparison. His defective emulative and comparative strategies extinguish his self. In sharp contrast to Carrie, Hurstwood tends to look backward rather than forward; he compares himself not to a better man he would like to be, but to a past self; he fears painful comparisons rather than seeking them out; and he shuts out any stimulus to change.

Hurstwood consistently resists all opportunities for comparison and emulation that would propel him to reconstruct a reputable or even a viable self. Unlike Carrie, who is a quick student of imitation, Hurstwood's "was not the order of nature to trouble for something better, unless the better was immediately and sharply contrasted" (*SC,* 86). He remains "unconscious of the marked contrasts which Carrie had observed" (*SC,*

---

38. Visiting Chicago four years after the publication of *Sister Carrie,* Henry James also notes the city "bristles" and orients itself to the future: "New York, and above all Chicago, were only, and most precariously, on the way to it, and indeed, having started too late, would probably never arrive" (*The American Scene,* 275, 278).

287). Hurstwood's lack of response to stimulating contrasts leads him to try to restrain others' desires. He tries to keep his daughter, Jessica, from buying new clothes (*SC,* 87-88) and from "display[ing]" herself (*SC,* 137), much as he will do later with Carrie, who is mistaken at least once for his daughter (Penn *SC,* 214). Soon after arriving in New York, Hurstwood fails to "delight" in Carrie's "expanding" desire: "He did not enthuse over the purchases. This led her to believe that neglect was creeping in," the narrator remarks in one of the most powerfully understated moments in the novel (*SC,* 294). Hurstwood, suggesting they move to smaller and smaller flats, "saw nothing remarkable in asking her to come down lower" (*SC,* 400). The one-time manager of appearances has already forgotten their necessity for sustaining a reputable self.

Especially in the Pennsylvania edition, Dreiser implies in many narrative interpolations that gender contributes to the difference in Carrie's and Hurstwood's behaviors. For instance, "Ah, how rapidly women learn.... Endow them with beauty, and within the possibilities of their environment they will pick and choose. Show them two men and they will understand which one appreciates women most. Such fine methods of comparison man does not possess" (Penn *SC,* 93). The narrator also remarks:

> In the progress of such minds [as Carrie's] environment is a subtle, persuasive control. It works hand in hand with desire. For instance, by certain conditions which her intellect was scarcely able to control, she was pushed into a situation where for the first time she could see a strikingly different way of living from her own. Fine clothes, rich foods, superior residence, a conspicuously apparent assumption of position in others,—these she saw. She was no more clever in observing this than any shop girl. No matter how dull is the perception in other things, in such matters all women are clear. (Penn *SC,* 97)

For Veblen, of course, modern women are better at invidious and emulative behaviors because it is their job to advertise their fathers' or husbands' wealth.

Perhaps discomfort with the prospect of constructing an advertisement for himself leads Hurstwood to shun visibility, a tendency again more pronounced in the Pennsylvania edition. It makes sense that when he initially faces his new identity as a criminal in Montreal, Hurstwood would seek out "inconspicuous place[s]," but his demeanor has tended to self-effacement for some time (Penn *SC*, 294). Even at Carrie's Chicago debut, Hurstwood made his "personality as inconspicuous as possible" (Penn *SC*, 113). Before he gets on the train for the fateful trip to Montreal, Hurstwood feels "[t]hat dread yearning for the fixed, the stable, the accustomed" (Penn *SC*, 288). As the narrator sums up Hurstwood's problem, he "took his situation too philosophically. He was too well satisfied" (Penn *SC*, 360).

Veblen contends, "It is much more difficult to recede from a scale of expenditure once adopted than it is to extend the accustomed scale in response to an accession of wealth" (*TLC*, 102). Hurstwood illustrates that point to tragic perfection, as he strives to remain "as nearly like the old Hurstwood" as he can, buying a nice meal, cigars, and otherwise "liv[ing] like a gentleman" (*SC*, 340, 341). As Veblen notes, some needs are more compelling than those necessary for survival, and so people do not "yiel[d] so abjectly before the pressure of physical want as to deny themselves all gratification of this higher or spiritual need" to maintain their "pecuniary decency" (*TLC*, 85). Hurstwood's "gentlemanly" spree after losing sixty dollars at poker reveals how easily "the means of comfort is diverted to the purpose of maintaining a decent appearance, or even a show of luxury" (*POS*, 392). Indeed, says Veblen, "[t]he last items of this category of consumption [conspicuous consumption] are not given up except under stress of the direst necessity. Very much of squalor or discomfort will be endured before the last trinket or the last pretense of pecuniary decency is put away" (*TLC*, 85). Again Veblen directs attention away from the act of consumption per se to ask *why* a person desires a particular object. Hurstwood's hanging on to a defunct reputation—the narrator terms it "the game of a desperate man" (*SC*, 365)—illustrates Dreiser's understanding that we spend most of our money not on necessities but "to keep us amused and socially important and to prevent our being laughed at" (*NL*, 186).

But the narrative voice in *Sister Carrie* also plays the desperate game. Not only does Dreiser repeatedly compare Hurstwood's failure with Carrie's success, he also invidiously compares the present George Hurstwood with his past self. The repeated designation, "ex-manager" (*SC*, 263), keeps alive the specter of Hurstwood's former self. Even before removing him from Chicago, Dreiser begins to force us to invidiously compare Hurstwood to an ideal of success clearly beyond his reach. After describing his position with Fitzgerald and Moy's, the narrator dismisses it as "greatness in a way, small as it was" (*SC*, 169).

In the pivotal chapter describing Hurstwood's regression in New York, the narrator demands that we, too, invidiously compare: "Walk among the magnificent residences, the splendid equipages, the gilded shops, restaurants, resorts of all kinds; scent the flowers, the silks[,] the wines; . . . and you shall know of what is the atmosphere of the high and mighty" (*SC,* 273). After evoking this dazzling display, the narrator shifts to Hurstwood's perspective: "here gathered all that he most respected on this earth—wealth, place, and fame. . . . [T]his show of fine clothes, place, and power took on peculiar significance. It was emphasized by contrast with his own distressing state" (*SC,* 274). Readers are urged to join Hurstwood, Carrie, and now the narrator in comparing not only success to suicide, but also manager to bum. Given Dreiser's identification with Hurstwood, the following confession in *Newspaper Days* is telling: "I never tired of looking at the hot, hungry, weary slums, any more than at the glories of the mansions of the west end [of St. Louis]. Both had their lure, their charm; one because it was a state worse than my own, the other because it was a better" (*ND,* 310). Dreiser claims to have been "afflicted" his entire life with the consciousness of "social misery," always seeing "[t]he rich districts as opposed to the poor ones," and on a smaller scale, "[t]he fine houses as opposed to the wretched ones." He concludes that "[t]he gulf was too wide, the comparisons cruel and unnecessary" (*DLR,* 252). Yet comparisons, however cruel, prove necessary to Dreiser in his first novel; he creates an authorial space for himself by such invidious comparisons between Carrie and Hurstwood. The heart, as Dreiser says, understands when confronted with contrasts (*SC,* 304).

Dreiser claims that as a youth, his brother Ed "used to rebuke me . . . for my sniveling desire to avoid publicity. Public opinion, 'what people will say,' exercised a strong influence upon me," an influence he claims later to have outgrown (*D,* 91). Dreiser's persona as a cultural critic, as we have seen, was based largely on defying public opinion, which suggests extreme self-consciousness about how he looked to others. But Dreiser's novels continue to illustrate Veblen's point that not only one's "good name" but also one's "self-complacency" depend on being observed with the things we own (*TLC,* 29, 37). Veblen clarifies the idea with his finely tuned irony: "When we say a man is 'worth' so many dollars, the expression does not convey the idea that moral or other personal excellence is to be measured in terms of money, but it does very distinctly convey the idea that the fact of his possessing many dollars is very much to his credit" (*POS,* 393). The reputable self is constructed not only by its alignment with objects; it must also be observed with these objects.

The Macroeconomic View: Invidious Comparison in *An American Tragedy*

The psychological model Dreiser uses in *An American Tragedy* closely resembles that of *Sister Carrie,* yet his perspective on it becomes far more critical. Clyde Griffiths is no one without fancy clothes; he remains no one if Gilbert, Sondra, and others fail to see him and confirm who he is. The central irony of the novel is that being caught in the act is embedded in Clyde's sense of self.

Dreiser found in the historical model for *An American Tragedy,* the murder of Grace ("Billie") Brown by Chester Gillette in 1906, "an entire misunderstanding, or . . . non-apprehension, of the conditions or circumstances surrounding the victims . . . *before* the murder was committed," and uses his novel to clarify the underlying causes (*SUP,* 296). Comparison and emulation provide the fundamental design for the novel. What fascinated Dreiser was less the murder of Brown than its representative nature. That the story keeps repeating itself becomes a pivotal fact within Dreiser's novel. Mirroring the author's compilation of factual accounts in preparation for *An American Tragedy,* Clyde Griffiths reads of a like incident in the *Albany Times-Union,* giving him the idea for his American tragedy. Entirely lacking an autonomous self, Clyde emulates the newspaper's "ACCIDENTAL DOUBLE TRAGEDY" so completely that intention becomes accident, and another double tragedy ensues (*AAT,* 438).[39]

Through Clyde, Dreiser explores the desire to be better than one's self that results from pecuniary emulation and invidious comparison. *An American Tragedy* provides a more macroeconomic look at this method of self-construction than does *Sister Carrie.* As James T. Farrell notes, the later novel features "greater social stratification" and "less movement from class to class"—although the *myth* of social mobility

---

39. On the factual basis of *An American Tragedy,* see Craig Brandon, *Murder in the Adirondacks: "An American Tragedy" Revisited;* Fishkin, *From Fact to Fiction;* and Susan Mizruchi, *The Power of Historical Knowledge.*

remains essential to the plot. Farrell links Dreiser's change of focus with two of Veblen's key terms, noting a "greater emphasis on consumption and leisure in *An American Tragedy.*"[40] Painting on a broader canvas, Dreiser focuses on the power of institutions to shape not only desires but the very self that possesses these desires. Because the constantly evolving community shapes desire and desire constructs the self, Clyde Griffiths's self is not only mutable but derives from what people around him dictate that he should want—even what he should be. When we meet Clyde in Book 1 he is learning how to construct an increasingly reputable self, but by Book 3 he ends up with an identity constructed entirely by others and, not surprisingly, he is so uncertain of who he is that he doesn't even know if he has murdered his lover.

An American Tragedy explores what Veblen describes as the "formal expression of . . . invidious comparison," or "the régime of status" (*TLC,* 291, 290).[41] The disparity between the haves and the have nots forms Dreiser's subject here, and so social class becomes even more an issue in *An American Tragedy* than in *Sister Carrie.* Clyde Griffiths, a product of what Veblen calls "vague and transient" class boundaries, shows how "the members of each stratum accept as their ideal of decency the scheme of life . . . in the next higher stratum" (*TLC,* 84). Although many readings of *An American Tragedy* trace the influence of Clyde's nuclear family—an influence that can scarcely be overemphasized—the standards of the wealthy Lycurgus Griffithses also shape Clyde's desires from the beginning of the novel, well before he meets them. As the narrator remarks in the second chapter, "The one thing that really interested [Clyde] in connection with his parents" was his "plainly different" Uncle Samuel (*AAT,* 17). Dreiser uses this plain difference between the two branches of the Griffiths family to explore the socially divisive and psychologically destructive effects of invidious comparison

40. James T. Farrell, "Theodore Dreiser," 161.
41. In "Some Neglected Points," Veblen refers to the regime of status in its more familiar meaning, citing Sir Henry Maine's two categories of social systems: "the system of status or the system of contract" (*POS,* 401). Technically, according to Maine's schema (followed by numerous thinkers, including Herbert Spencer), early twentieth-century American society would be characterized by contract, not status. But Veblen's use of *status* to describe his contemporary social organization was not anachronistic; he deliberately deploys an existing social scientific term against the grain (see, for instance, *Leisure Class,* 291, 292). Veblen observed scant indication of truly free contract for modern "class distinctions, except pecuniary distinctions, have fallen away," leaving two basic categories: "those who have more and those who have less" (*NOP,* 151).

and pecuniary emulation. In the process, Dreiser shows that Clyde's self is a thing as surely fabricated, and as disposable, as the shirt collars produced in his uncle's factory.

Besides the people Dreiser calls "asinine," the small fry who look up to Rockefeller and Morgan, the rich themselves also practice invidious comparison (*SUP,* 269). As Veblen puts it, "[t]he leisure class stands at the head of the social structure in point of reputability; and its manner of life and its standard of worth therefore afford the norm of reputability for the community." The leisure class is both the origin and the final goal of all pecuniary emulation. As class lines blur, the "coercive influence" of leisure-class standards of pecuniary reputability spreads more rapidly (*TLC,* 84).

In a retrospective piece for *Mystery Magazine* called "I Find the Real American Tragedy" (1935), Dreiser notes the existence of "our 'leisure class,' the Four Hundred of New York and the slave aristocracy of the South," a group "not so much interested in work or mental or national development . . . as they were in leisure and show." Dreiser's mentioning the group's opposition to work and interest in display—to say nothing of his highlighting the phrase "leisure class"—suggests he has Veblen's definition in mind. Dreiser further notes that "the ambition to be somebody financially and socially" is not "merely the attainment of comfort . . . , but rather the accumulation of wealth implying power, social superiority, even social domination," in other words, invidious distinction and power (*SUP,* 292, 291).

Dreiser mentions in the same essay that the leisure class has its "imitators" (*SUP,* 292). The imitators in *An American Tragedy,* although they pass in the community for the real thing, are the Lycurgus Griffithses.[42] Mrs. Samuel Griffiths even looks down on the Finchleys—a perspective absolutely unfathomable to Clyde, who can imagine no family more exalted. Whereas Veblen sees "chronic dissatisfaction" as typical of the "normal, average individual," he notes that a member of the highest class will demonstrate "a restless straining to place a wider and ever-widening pecuniary interval between himself and this average standard" (*TLC,* 31). It is an inevitable consequence of invidious comparison that the Lycurgus Griffithses need to maintain a ceremonial distance

---

42. On the far-reaching significance of authenticity and imitation in late nineteenth- and early twentieth-century American culture, see Miles Orvell, *The Real Thing: Imitation and Authenticity in American Culture, 1880–1940.*

from their poor relation to confirm their superiority, and herein lies the seed of the tragedy: the Griffithses cannot behave consistently toward Clyde, much less become the new family he desires.

Because he is a Griffiths, Clyde is brought to Lycurgus to work, yet he soon discovers some Griffithses to be more equal than others. Bluntly enough, Samuel places his nephew in the "very bottom of the business . . . —the basement of the . . . plant." (His rationalization, that Clyde "was to be taught the business from top to bottom," is curious indeed in light of the facts [*AAT,* 175].) But again this treatment, which no doubt confuses Clyde, is inevitable. As Veblen explains, the segregation of "worthy" from "unworthy" employments sustains what he terms the honorific (leisure) class (*TLC,* 8). Without a lower class, the upper crust would have no one to be superior to. To maintain their social standing, the Lycurgus Griffithses need to keep their distance from Clyde, just as Clyde must look ever upward to improve his own.

The Lycurgus Griffithses also must actively police their social status, as Clyde's aunt shows with her social distinctions between her own family and the "manufacturers of bacon, canning jars, vacuum cleaners, wooden and wicker ware, and typewriters"—to say nothing of the frequently discussed differences between the Lycurgus and Kansas City branches of the family (*AAT,* 148). The Lycurgus Griffithses are proud of their Wykeagy Avenue home precisely because of the distance it puts between them and others, because they "had been so long climbing up to it" (*AAT,* 215). That the social distinction between Finchleys and Griffithses is so fine as to prove invisible to the observant Clyde only confirms its value; the truly elite, Veblen observes, discover subtle ways to display their wealth to distinguish themselves from more conspicuous arrivistes (*ECO,* 70-71; *TLC,* 235).

Samuel Griffiths's means for sustaining his self-image sheds light on what happens to Clyde, for the nephew emulates—even anticipates—his uncle. Apart from rare moments demanding active intervention to preserve his status, Samuel can placidly accept "himself at the value that others placed upon him" (*AAT,* 153). As Veblen says, "the usual basis of self-respect is the respect accorded by one's neighbours" (*TLC,* 30). But Samuel can confidently leave his reputation for others to confirm because his pecuniary standing is the highest in Lycurgus, a fact decidedly not true for his nephew. The Lycurgus patriarch correctly sees a social chasm separating him from Clyde and cautions his family, "I don't expect any of you to pay [Clyde] any social attention—not the slightest." Samuel reasons (as it turns out, incorrectly) that Clyde is "not the sort of boy anyhow, that would want to put himself on us" (*AAT,* 158). It seems strange the factory owner would not know that constructing a successful self means precisely the imaginative putting of oneself on at a higher level.

Following the most successful man he knows, Clyde accepts his self based on others' evaluations. Because these evaluations are inconsistent and contradictory, the results are tragic. The importance of visibility and reputation in determining self-esteem is so ingrained in Clyde that he has always seemed most authentic to himself when reflected in another's eyes. Before leaving the Kansas City mission, he formulates the grounds of his objection to his parents' work: it "was not satisfactory to others" (*AAT,* 14). After he has begun hiding money from his mother and learns of his sister's pregnancy, the trend of Clyde's personality is determined when Esta "looked at him admiringly and he was properly impressed by her notice of him" (*AAT,* 96). This method of securing self-esteem by relying on others' favorable notice becomes Clyde's modus operandi. Once in Lycurgus, he briefly constructs a reputable self by looking back imaginatively on where he had been so as to generate a favorable contrast: "What would Ratterer think if he could see him now—his uncle's great house and factory?" (*AAT,* 190). With the mental image of an old friend's opinion validating his present self, Clyde finds it easier to pursue his desires. For instance, he convinces Roberta to get a private room for their assignations while "[c]arried away by a bravado which was three-fourths her conception of him as a member of the Lycurgus upper crust" (*AAT,* 280). Although both Carrie and Hurstwood care about how they look to others, Clyde's self all but dissolves in the speculations (in both the cognitive and visual meanings of that word) and assumptions of others.

Well before meeting the Lycurgus Griffithses, Clyde tries to validate himself as they do. Invidious comparison, demanding the maintenance of a ceremonial distance from those below, leads Clyde to attempt to construct a self by looking down. His sense of superiority goes further than Hurstwood's—and rests on facts just as dubious. An early description of Clyde notes he was "as vain and proud as he was poor," and always "looked upon himself as a thing apart" (*AAT,* 18). Clyde's contempt for the homeless who frequent his parents' mission— "loafers, drunkards, wastrels, the botched and helpless" (*AAT,* 17)— magnifies as he learns to look down on the speech and manners of the bellboys he works with at the Green-Davidson Hotel (*AAT,* 34, 38), a onetime friend in Lycurgus (*AAT,* 196), and, most ominously, Roberta: "For after all, who was she? A factory girl!" (*AAT,* 301). Yet who was he? The narrator describes a torn Clyde, working at the Green-Davidson and pursuing Hortense Briggs, as "part and parcel of this group . . . he looked down upon" (*AAT,* 72). Although part of him struggles to earn acceptance in a group, another part finds validation only by excluding others, as do those with higher status claims.

Many characters mistake Clyde's condescension for the superiority it is intended to resemble and reinforce a sense of self that he cannot sustain. The Kansas City prostitute to whom Clyde loses his virginity fans his superiority by saying, "You're not like those other fellows. You're more refined, kinda" (*AAT,* 68). By Book 2 he has adopted a "conscious gentility of manner. . . . considerably different from the Clyde who had crept away from Kansas City in a box car" and can present to his uncle in the Chicago Union League Club "a somewhat modified version of the one [the Clyde] who had fled from Kansas City" (*AAT,* 160, 159). Carrie Meeber/Wheeler/Madenda anticipated Clyde's proliferating selves, but Carrie consciously set out to become an actress. Clyde does not realize that he too is acting. His new self needs ever more convincing validation, and secretaries, petty managers, and others at the Lycurgus factory respond as if on cue. Confirming Clyde's belief that "[o]bviously it was no small thing to be a Griffiths here [in Lycurgus]" (*AAT,* 186), he receives "unusual deference" (*AAT,* 179). Clyde is only doing what comes socially. As Veblen remarks, "Regard for one's reputation means . . . emulation. It is a striving to be, and more immediately to be thought to be, better than one's neighbor" (*POS,* 392).

One danger with this method of self-construction arises from the fine line between emulation and dissimulation. Veblen says that "the appearance of success is very much to be desired, and is even in many cases preferred to the substance" (*POS,* 394). The appearance of success is exactly what Clyde gains, if not in his own eyes (if there is such a thing apart from what he sees reflected in others' eyes), then in the eyes of Roberta, Sondra, and Lycurgus citizens in general. But Clyde's Lycurgus identity is based on a series of mistakes about who he is. What he takes to be the redemptive event of his life, Sondra's interest in him, would never occur were he still the Clyde of Kansas City (and isn't he?) or were it not for his striking physical resemblance to Gilbert, for whom Sondra mistakes Clyde the first time she sees him. Even Roberta, who like Clyde was "always thinking of something better," is drawn to his supposed, but illusory, status (*AAT,* 245). It is madness to say that how one *looks* is all that matters, but that premise of the conspicuous modern self is exactly what Dreiser seizes on, and critiques, in *An American Tragedy*. In doing so, Dreiser shows how a simulation comes to seem more real than reality.

Clyde's markedly irrational behaviors, especially during moments of stress, such as his hallucinations while out on the boat with Roberta, demonstrate his instability. It has been well established that Dreiser drew on Freud, whom he had read during his years living in Greenwich Village (particularly around 1918), in the construction of Clyde's psyche. But the particularities of Clyde's irrational behaviors also provide a good illustration of Louis Schneider's contention that Freud's theory of neurosis overlaps considerably with Veblen's psychological model. Schneider's description of the Freud-Veblen nexus seems made to order for Clyde: "The basic insecurity and doubt about one's worth . . . make it understandable that the 'neurotic' should ceaselessly strive to manifest superiority. Since the doubt can never be put to rest. . . . [one consequence is] the fundamental characteristic of not being 'one's self' or of being 'self-alienated.'" It is difficult to imagine a character more self-alienated than Clyde becomes in Book 3.[43]

Clyde also illustrates, with a depressing degree of accuracy, the "narcissistic personality" as identified by Christopher Lasch. According to Lasch, the modern narcissist is the product of the shift from a production-oriented economy to a consumption-driven one. One result of this economic shift is precisely what Veblen notes: in Lasch's words, "Success . . . has to be ratified by publicity." Possessions matter less than appearances, and so the narcissist must turn to others to validate who he is. "He cannot live without an admiring audience," remarks Lasch, a comment that applies to Carrie and to Hurstwood almost as readily as to Clyde. (In Clyde's case, he cannot live with an admiring audience, either.) Having scant inner resources, the narcissist exists "in a state of restless, perpetually unsatisfied desire." Thus the psychology of desire—as illustrated by both Dreiser and Veblen—anticipates what Christopher Lasch critiques half a century later as "the culture of narcissism."[44]

43. Louis Schneider, *The Freudian Psychology and Veblen's Social Theory,* 199–200.
44. Lasch, *Culture,* 122, 117, 137, 38, 356, 23.

Combining in an unstable blend the forward-looking reflexes of Carrie with the backward-looking tendencies of Hurstwood, Clyde's thoughts bounce between the Lycurgus Griffithses and his own family or others from his past. He can scarcely be said ever to exist in the present, for he constantly compares himself to more reputable others and strives to improve himself by emulation. Like Carrie, Clyde's "desire for more—more—that intense desire" bypasses consciousness or will (*AAT*, 804). As a youth in Kansas City, "before he had ever earned any money at all, he had always told himself that if only he had a better collar, a nicer shirt, finer shoes, a good suit, a swell overcoat like some boys had! . . . And pretty girls with them. And he had nothing" (*AAT*, 19). "If only" might well be Clyde's motto, for he characteristically inhabits a hypothetical future, purchased on credit against funds he does not yet have.

Contrasts have always overpowered this unstable self. The Green-Davidson Hotel, contrasting sharply with "the timorous poverty" of Clyde's home, "was more arresting, quite, than anything he had seen before" and soon "caused him to think differently of how one should live" (*AAT*, 32, 92). Dreiser's quiet pun on the word *arresting* foreshadows the alignment of conspicuity with criminality to come. Just as Clyde's parents' tawdry mission makes the "gauche luxury" of the Green-Davidson seem tony to him, the hotel world makes Lycurgus's Wykeagy Avenue seem, again by contrast, stupendous. Even his nasty cousin Gilbert's condescension strikes Clyde favorably: "How wonderful . . . to . . . take oneself so seriously. . . . [T]his youth was very superior and indifferent . . . toward him. But think of being such a youth" (*AAT*, 218). Clyde longs to emulate Gilbert so that he can take himself seriously too.

The pivotal chapter 42 of Book 2, where Clyde conceives of the imitative murder that will prove most dangerous to the continuance of his self, Dreiser blatantly describes as a moment of invidious comparison. Fate itself seems bent on endorsing the contrast between Clyde's past and present desires as letters arrive "simultaneously" (*AAT*, 433) from Sondra and Roberta.[45] After reciting Sondra's letter, the narrator interrupts Roberta's letter to emphasize its "contrast" with Sondra's before sealing the plot by saying, "And it was the contrast presented by these

---

45. Dreiser later claims that Roberta's letter arrives in "almost the same mail, at least the same day" (*AAT*, 433).

two scenes [evoked by the letters] which finally determined for him the fact that he would never marry Roberta—never" (*AAT,* 435, 436).

Invidious Comparison and Pecuniary Emulation on Trial

Dreiser emphasizes in brutal detail the extent to which Clyde's self is constructed in the eyes of his beholders during the trial and punishment sequence, where his reputation prevails, and he becomes a spectacle. Clyde becomes, to all intents and purposes, an image of himself drawn from how others view him. Fredric Jameson analyzes a similar phenomenon, when he describes as the ultimate form of objectification, "precisely the image itself. With this universal commodification of our object world, the familiar accounts of the other-directedness of contemporary conspicuous consumption and of the secularization of our objects and activities are also given: the new model car is essentially an image for other people to have of us."[46] Images of Clyde proliferate during Book 3 of *Tragedy*. The trial is an orgy of commodification and conspicuousness; Clyde's "self," completely on display, literally enters the marketplace. Oglers, fed by the "picturesque accounts" as much as by the peanuts and hotdogs—all of which are for sale—find a "holiday or festival" atmosphere (*AAT,* 577, 630, 629). The regime of status prevails over individual identity as Clyde is systematically reconstructed into what he wishes he were but never will be: a wealthy Griffiths. As the assistant deputy "was inclined to respect" the "class" to which Clyde supposedly belongs, and as sheriff, jailers, and spectators show "sycophantic pride in his presence" (*AAT,* 555, 572), his alleged pecuniary strength is now written in letters so large that, as Veblen says, all who run may indeed read. As Thomas Strychacz astutely puts it, "Clyde's 'self' is not integral or original, but emerges from a matrix of superimposed texts"; he is "rewritten" throughout the novel based on various accounts in the mass media.[47]

During the trial, Clyde is destroyed by his illusory status and the invidious comparison and pecuniary emulation performed by those around him. District Attorney Orville Mason sets the tone for the prosecution when he invidiously compares Clyde to Roberta, which

---

46. Fredric Jameson, "Reification and Utopia in Mass Culture," 132.
47. Strychacz, *Modernism,* 94.

the narrator reveals to be motivated by Mason's misperception of an invidious distinction between himself and Clyde: "Wealth! Position! . . . For the social difference between this man and this girl from his point of view seemed great. . . . [W]as she not poor? . . . And was that not part and parcel of a rich and sophisticated youth's attitude toward a poor girl? By reason of his own early buffetings . . . the idea appealed to him intensely. The wretched rich!" (*AAT,* 515-16). The linchpin of Mason's prosecution will be this sort of invidious comparison. The trial pivots on discerning the identity of Clyde's self—in Mason's words, "who is the individual" we recognize as Clyde Griffiths (*AAT,* 641). The ambiguity derives from the discrepancy elaborated in Book 2 between "who the Griffiths were here [in Lycurgus], as opposed to 'who' the Griffiths were in Kansas City, say—or Denver" (*AAT,* 189). With masterful irony, Dreiser has Mason ask rhetorically, "Is he the son of wastrel parents . . . ? Is he?" (*AAT,* 641). Telling the jury that Clyde "had more social and educational advantages than any one of you," Mason encourages them to invidiously compare. The jurors and citizens do compare, with a vengeance.

Even Clyde's defense attorney, Reuben Jephson, introduces invidious comparison into the courtroom when he asks his client to explain to the jury "just what it was about this Miss X, as contrasted with Miss Alden" (*AAT,* 685). Clyde cannot answer that crucial question coherently, for Sondra simply was all that he did not have, all that was out of his reach, all that he longed to be. At first sight she seemed to him "so different . . . so superior" (*AAT,* 219) to any girl he had ever seen. Her invidious superiority attracted Clyde all along. Indeed, the erotic attraction that once pulled Clyde so powerfully to Roberta has been virtually absent in his dealings with Sondra—a striking omission in the psychology of a Dreiser male, and one that deserves scrutiny. According to one of Veblen's campier theories, the desire for private property started when early *Homo sapiens* captured women from competing tribes. The seized women were seen not as sexual objects but as "trophies" to display (*TLC,* 23; see also *ECO,* 32-49). Similarly, Sondra seems to Clyde the ultimate trophy, arousing in him not sexual passion but the pangs of emulation: "a curiously stinging sense of what it was to want and not to have" (*AAT,* 220).

Dreiser's cultural criticism in *An American Tragedy* becomes particularly sharp when he turns to examine the institutions purporting to serve justice. He provides a much harsher evaluation of invidious comparison and pecuniary emulation here than in *Sister Carrie,* as evidenced by the fact that Hurstwood declines all the way to suicide because he fails at this method of self-construction, whereas Clyde goes to the electric chair for being so good at it. Clyde's success at constructing himself in the socially approved manner becomes his tragedy. As Dreiser remarks, Clyde's "was not an *anti-social* dream as Americans should see it, but rather a *pro-social* dream" (*SUP,* 297). The narrative voice in *Sister Carrie* participates in the logic of invidious comparison, but in Book 3 of *An American Tragedy*, the narrator *describes* invidious comparison without participating in it; he no longer, to use the distinction Georg Lukács discusses in a different context, *narrates* invidious comparison.[48]

Dreiser's criticism of this method of self-construction points unambiguously to the surrounding culture. While jury, lawyers, and citizens of Catarqui County become increasingly confident that they have determined just "who" Clyde is, the narrator becomes less so. Thus the sympathetic reader is left uncertain about where to place blame for the central act in the novel: Who killed Roberta? Did Clyde? Or did the "America" of the novel's title, and if so, what does that mean? Clyde goes to his death unsure if he has committed murder; readers can debate for hours "what happened" on the rowboat (as classroom discussions readily confirm); only for the community that has constructed him does Clyde Griffiths become a fixed entity: exactly what they want him to be.

The self-proclaimed success of the legal system in determining who Clyde is (a rich, irresponsible, oversexed murderer) illustrates how the canons of emulation and comparison can leave one's self a product of institutional thinking. But however much a "mental as well as a moral coward" (*AAT,* 669) or even criminally negligent Clyde may be, he does not, in fact, have the social status or money that incites the jury to convict him of murder. Even if the jury was right in its decision about Clyde's guilt—and Dreiser's novel strikes me as too ambiguous to permit such a reading—it would be right for the wrong reasons.

Veblen explains that "[a] culture whose institutions are a framework of invidious comparisons" will only accept conclusions that reinforce

---

48. In "Narrate or Describe?" Georg Lukács uses these terms to distinguish literary realism from naturalism. My interest is not in labeling *An American Tragedy* "naturalist" as opposed to "realist" (or vice versa), but Lukács's distinction is provocative nonetheless.

what it already believes. Such a culture demands "truth . . . of a ceremonial nature" which it treats like "reality regardless of fact" (*POS,* 107-9). Throughout Book 3, Dreiser stresses almost beyond the limits of human endurance the distance between ceremonial reality and fact. He shares Veblen's skepticism about the accessibility of ultimate truth. Indeed, one of the triumphs of Dreiser's masterpiece in the "realist" tradition is its calling into question so many of the supposed foundational beliefs of literary realism, such as the allegiance to factual accuracy and objectively drawn detail. (In one of the most blatant examples of Dreiser's undermining the authority of facts, the assistant district attorney plants Roberta's hairs in Clyde's camera, concocting pseudo-evidence of murder.) Documenting American culture's obsession with reputation and appearance, Dreiser critiques the notion that reality is visible or even accessible to human consciousness.

A line from a recent song by David Byrne sums up the psychological model I have been examining: "I am just an advertisement for a version of myself."[49] Dreiser was personally familiar with the method of self-construction that accounts for Carrie's success on stage, for Hurstwood's suicide, and for Clyde's death sentence. In his autobiographical account of his professionalization, *A Book about Myself,* he repeatedly registers the importance of how the self looks to the eyes of others. For instance, Dreiser recalls the first time he was taken seriously as a poet by colleagues while working for the *St. Louis Globe-Democrat* in 1892. "To be considered a writer" by others "raised me in my own estimation," he records (*BM,* 129). But a year later, traveling to Grand Rapids, Ohio, on one of those "mobile palaces," a Pullman, Dreiser felt himself a "failure. Other men had money." The Drouet-style "drummers" in the train, in particular, "seemed . . . the most fortunate of men," and so Dreiser decided to follow their lead: "I assured myself that the way to establish my true worth was to make everyone else feel small by comparison" (*BM,* 361-62).

Dreiser's first fictional masterpiece, a bildungsroman of invidious comparison, insightfully displays this method of self-construction through the fates of Carrie and Hurstwood, at times even endorsing it by the narrator's own invidious comparisons. Dreiser's last fictional

---

49. David Byrne, "Angels."

masterpiece demonstrates the reductio ad absurdum of invidious comparison, for Clyde's identity becomes mysterious to him and inscrutable to readers. But the fact that Carrie, Hurstwood, Clyde, and other Dreiser characters gain no more than a "fraction" of their "original desires" (*SC,* 462) reflects less on the individuals than on the institutions constructing them. For Dreiser as for Veblen, the watching, comparing, emulating self-in-construction replaces the self-contained globule of desire. Pecuniary emulation and invidious comparison eradicate the notion of intrinsic personal worth or an autonomous self. "Remove contrast," asks Dreiser, "and what have you?" (*NL,* 328).

4

Gender and Cultural Criticism

The Instinct of Workmanship in *Jennie Gerhardt*

While Dreiser's best-known novels exemplify central ideas in Veblen's social and economic theory, the judgment Dreiser passes on these subjects—such as on the immaterial basis of business illustrated in the *Trilogy of Desire,* or on the psychological model of invidious comparison and pecuniary emulation seen in *Sister Carrie* and *An American Tragedy*—can differ dramatically from Veblen's. This chapter takes up the area of most complete axiological congruity between Veblen's and Dreiser's thinking, highlighting their cultural criticism in a particularly sharp, if idiosyncratic, form.

As Rick Tilman demonstrates, most commentators have found the ethical, political, and philosophical values underwriting Veblen's criticism of American culture difficult to pin down. One reason for the uncertainty is the elusiveness of Veblen's satire; another is that he spends far more time anatomizing what is wrong than providing alternative models, either for social organization or for individual behavior. But Veblen's values are clearly evident throughout his writings if one looks in the right place. Notwithstanding his contempt for invidious comparisons, Veblen's entire social theory is structured on binary oppositions, and his own values always reside in the term of any dichotomy that he says his culture despises. For instance, he prefers industry over business, utility over waste, substance over ceremony, the engineer over the financier, technology over salesmanship—and through such invidious comparisons, Veblen registers opposition to what he claims are dominant American values. The ethical glue holding Veblen's entire analytical matrix together is his belief that most Americans value contemptible things and despise worthy ones. This belief grounds Veblen's cultural criticism and correlates with his insistence that intellectual work must be conducted on the margins of society.[1]

In *Jennie Gerhardt* (1911), Dreiser stakes out the same ethical, political, and philosophical position. He admires the title character (indeed, he idolizes her) for her indigence, her kindness, her generosity, and Dreiser depicts these qualities as helping Jennie to maintain her integrity in the face of institutional pressures and ceremonial considerations. Dreiser also insists that other characters look down upon Jennie for precisely these reasons. Through the systematic contrast of Jennie with the culture in which she lives, he indicts the "marvelously warped" and "radically wrong" values that are socially dominant (*JG*, 92, 93).[2] The following passage encapsulates the logic of the novel:

> The world into which Jennie was thus unduly sent forth was that in which virtue has always struggled. . . . Virtue is that quality of generosity which offers itself willingly for service to others, and, being this, it is held by society to be nearly worthless. Sell yourself cheaply and you shall be used lightly and trampled under foot. Hold yourself dearly, however unworthily, and it will come about that you will be respected. . . . [Society's] one criterion is the opinion of others. Its one test, that of self-preservation. (*JG*, 87)

The definitive contrast here is between worth (actual, innate) and value (illusory, socially determined). The novel generates multiple contrasts between Jennie's worth, as seen especially in her philosophy of "virtue" and "service," and the pecuniary accountancy of those around her. Following through on this complex judgment that so closely parallels Veblen's, in *Jennie Gerhardt,* Dreiser sustains one of his strongest pieces of cultural criticism.

1. Tilman, *Thorstein Veblen and His Critics.* On Veblen's social theory as structured on invidious comparisons, see my "Veblen's Anti-Anti-Feminism."
2. I use the Pennsylvania edition published in 1992 unless otherwise noted, because both Jennie's character and Dreiser's cultural criticism are better defined here. In 1911, Harpers agreed to publish *Jennie Gerhardt* only under the condition that controversial material be cut. The result was that, in the words of James L. W. West III, "*Jennie Gerhardt* was transformed from a blunt, carefully documented piece of social analysis to a love story merely set against a social background" ("The Composition and Publication of *Jennie Gerhardt,*" in *Jennie Gerhardt,* 442).

Readers have traditionally seen *Jennie Gerhardt* as Dreiser's "softest" novel, largely for two reasons involving the title character. First, Jennie is not the unmistakable product of modern, capitalist America for which Dreiser is best known—such as the wistful starlet, the "ex-manager," the artist-adman, the financial tycoon, the youth whose belief in the American dream leads to his tragic end, the pregnant working girl who dies trying to marry her way to respectability. Jennie is neither ambitious nor invidious, she enjoys working (particularly in service to others) and is content to wear a simple dress. But the qualities making her unusual in Dreiser's fiction are not signs that the author has faltered. In fact, the narrator goes out of his way to stress Jennie's uniqueness with interpolations such as this: "Caged in the world of the material, . . . such a nature [as Jennie's] is almost invariably an anomaly" (*JG*, 16). One of Dreiser's generalizations helps to clarify what sets Jennie apart from most of his other characters: "We live in an age in which the impact of materialized forces is well-nigh irresistible; the spiritual nature is overwhelmed by the shock" (*JG*, 125). So do most of Dreiser's characters discover—and so would Veblen say is true for most United States citizens. But Jennie resists; her spirit is not overwhelmed. Yet since the qualities distinguishing her from Carrie Meeber, Clyde Griffiths, Roberta Alden, George Hurstwood, and Frank Cowperwood make it difficult to interpret Jennie according to the concepts familiar to most theories of literary realism and naturalism, many Dreiser critics choose conveniently to ignore or to pillory *Jennie Gerhardt*.

Historically, theories about literary realism and naturalism have stressed the idea that individuals are products of their environments. Realist and naturalist novels, it has often been asserted, weigh in heavily on the side of social construction: these characters are made, not born.³ Particularly on this count, *Jennie Gerhardt* is a problematic text, which brings us to the second reason for its relative critical neglect: Dreiser

---

3. A notable exception would be naturalist novels that illustrate racial or ethnic determinism—Frank Norris's *McTeague*, for instance, although even here the characters operate according to social as well as biological "forces." My point is not that naturalism should be equated with *any* particular model of characterization; indeed, I find the whole pigeonholing process dubious.

A range of treatments of literary naturalism as deterministic can be found in Malcolm Cowley, "A Natural History of American Naturalism"; Walcutt, *American Literary Naturalism;* Philip Rahv, "Notes on the Decline of Naturalism"; Pizer, *Novels;* John J. Conder, *Naturalism in American Fiction: The Classic Phase;* Martin, *American Literature;* Mitchell, *Determined Fictions;* and Howard, *Form and History.*

advances a decidedly essentialist conception of womanhood that many readers find distasteful, and understandably so. Given Dreiser's own well-known sexual "varietism," his apostrophes to Jennie's yielding and giving nature can seem self-serving, at times downright embarrassing.

Take, for instance, the "service" he keeps insisting upon, and that critics from H. L. Mencken on have singled out as one of Jennie's defining characteristics. Even after her second lover withdraws into condescending righteousness upon the discovery of Jennie's illegitimate daughter from an earlier liaison, she worries if "in this crisis there was some little service she might render [him]" (*JG*, 212). Dreiser asks us to believe that Jennie enjoys serving others, even (or especially) men who outrank her socially but who are not, morally speaking, her equals; furthermore, Dreiser suggests that her service ennobles her.

The nobility that Dreiser attaches to Jennie's "serviceable and harmonious" disposition corresponds to Veblen's approval of human behaviors that are other-directed (*JG*, 364). The concept of "serviceability" occurs throughout Veblen's writings, always given a favorable cast. As seen in his distinction between industry and business, Veblen frequently contrasts "serviceability" with "vendibility" (i.e., useful work versus salesmanship). And, like Dreiser, Veblen considers women innately more disposed to serviceability. Consider in this context how aptly Veblen's words characterize Jennie: "So long as the woman's place is consistently that of a drudge, she is, in the average of cases, fairly contented with her lot. She not only has something tangible and purposeful to do, but she has also no time or thought to spare for a rebellious assertion of such human propensity to self-direction as she has inherited" (*TLC*, 358–59). Despite the irony in this passage (or perhaps because of it), it is clear that Veblen values the "tangible and purposeful" work done by his hypothetical "drudge." Dreiser does not see Jennie in this light, although her resignation to let others determine her life follows Veblen's prediction. In contrast to Dreiser's other characters (particularly Carrie, Clyde, and in a different sense, Cowperwood) Jennie never rebels or asserts herself. Jennie does, however, maintain her integrity in a way that quietly, if unconsciously, serves to critique the self-seeking of those around her. As Dreiser depicts her, Jennie is more heroine than victim; she is, paradoxically, a strong woman because of her weakness.[4]

4. Until the collection of essays edited by James L. W. West III, *Dreiser's "Jennie Gerhardt": New Essays on the Restored Text*, the novel rarely received sustained

While I should not care to endorse this model of femininity, the gender essentialism should not block us from looking at the cultural work Dreiser's novel performs. As Diana Fuss argues, although essentialist logic has historically been used more often to confine women than to liberate them, the belief is not inherently (one might say, not essentially) reactionary or progressive. What matters, according to Fuss, is how gender essentialism is *used* in a particular context.[5] The cultural criticism embedded in *Jennie Gerhardt* becomes clear once we read Jennie's womanly "serviceability" in a positive light. Veblen's quasi-anthropological musings about what he calls the instinct of workmanship and the parental bent provide a context for such a reading. Like many feminists and protofeminists from the mid-nineteenth century through the early years of the twentieth, Veblen and Dreiser use essentialist views of women's work, which they align with motherhood, to criticize the values of capitalist America. And like many poststructural analysts, Dreiser and Veblen see social class, gender, and power as interlocking determinants.

The Ethical Center: Workmanship and Motherhood

First, some definitions. Although best known for his theory of leisure and for defining business as the antithesis of productive labor, Veblen also theorized extensively about work. (Indeed, his contempt for business is premised upon the existence of the realm of legitimate labor he labels industry.) Veblen intended to devote his second book to the subject of work, but *The Instinct of Workmanship and the State of the Industrial Arts*—said to be his favorite among his works—was not published until 1914. A "non-invidious" trait, workmanship demonstrates the "proclivity for purposeful action" that is a "generic feature of human nature" (*IOW,* 194; *ECO,* 80). The instinct of workmanship accounts for the human tendency to make things, useful things; its "functional content" is that trait so pronounced in Jennie Gerhardt—"serviceability" (*IOW,* 31). Workmanship is so important to Veblen's thinking that T. W. Adorno calls it his "supreme anthropological category," and Warner Berthoff notes that Veblen's

---

attention. Many of these essays argue in various ways for the strength of Jennie's character or philosophy.

5. Fuss, *Essentially Speaking.*

"celebration of the ...'instinct of workmanship' mythopoeically balances [his] world-historical pessimism." Nevertheless, as Christopher Shannon observes, "for all of the attention given to [Veblen's] attack on 'conspicuous consumption'..., little or no attention has been given to the idea of production—of 'workmanship'—that informs this critique."[6]

Likewise in *Jennie Gerhardt*, the title character's goodness, which Dreiser makes palpable through her devotion to work and service, balances the darker vision he advances elsewhere—even elsewhere in the same novel. In *Tales of the Working Girl*, Laura Hapke credits Dreiser for "elevat[ing] the working girl story into art," but she faults him for "censoring" Jennie's "involvement" in menial work. She notes that Dreiser does not give Jennie's work even the few pages of description he gives Carrie's—an acute observation, yet one that leads Hapke to give Dreiser less than his due. As Nancy Warner Barrineau, who reads Dreiser's depiction of women's work more sympathetically, observes, *Jennie Gerhardt* is in fact "much more radical" than *Sister Carrie*.[7]

From the first scene in *Jennie Gerhardt*, depicting two poor women seeking employment, the novel is permeated by Dreiser's admiration for women's work. Rather than censoring Jennie's labors, Dreiser saturates the novel with her instinct of workmanship. Left pregnant by Senator Brander, Jennie finds "the pleasure of work lifting her out of herself" (*JG*, 95). "I must work. I want to work," she responds when Lester asks her to become his mistress (*JG*, 133). The internal compulsion exists because Jennie is the "soul of industry"; "I like things to do," she explains (*JG*, 252). The narrator remarks several times that Jennie's "industry" is "natural" (197, 396). As he puts it, she has a "need of doing something, even though she did not need to"—work is a requirement for Jennie's personal fulfillment, even when she has no economic need (*JG*, 389). Thus Jennie holds a job outside the home for a time when she doesn't need the money (*JG*, 173). Simply disposed "against idleness," Jennie "liked to be employed" (*JG*, 389, 396). Although Lester eventually deserts her, Jennie's "natural industry" never does (*JG*, 197).

---

6. Dorfman, "New Light," 98; Adorno, "Veblen's Attack," 82–83; Berthoff, "Culture and Consciousness," 495; Shannon, *Conspicuous Criticism*, 3.
7. Laura Hapke, *Tales of the Working Girl: Wage-Earning Women in American Literature, 1890–1925*, 71, 82; Nancy Warner Barrineau, "'Housework Is Never Done': Domestic Labor in *Jennie Gerhardt*," 127.

Dreiser's appreciation for the domestic labors associated with "women's work" derives from his earliest memories. He understood well "the age-old slavery of domestic duties" (*DLR*, 167). In *Dawn* (1931), he tallies up the tasks his mother did to nurture and serve the family: "cleaning, washing, baking, preserving such fruits as she could get for little or nothing. . . . sewing and tailoring." The family's poverty forced Sarah Dreiser to extend her labors to doing other people's washing and ironing, even to taking in boarders (*D,* 65). Although Dreiser voices considerable embarrassment about his impoverished childhood (including the presence of boarders in his home [*D,* 98]), he reverences his mother for her strenuous domestic labors. Dreiser wistfully recalls, "I can see her now, sitting by an open window in summer, from which the fields presented an idyllic view, or near the fire or under a lamp in winter, stitching, stitching, stitching" (*D,* 65). As the repetition of the final word suggests, Sarah Dreiser's work was endless, yet the son recalls the setting, and the memory, as "idyllic." Dreiser says, "And yet with all these drawbacks, ours was a home, full of a kind of sweetness that never since has anywhere been equaled for me, however difficult it must have been for her" (*D,* 66). The alignment of domestic work with motherhood is hardly unusual; what is striking is Dreiser's attempt to reclaim the value of traditionally underrated work and to use it, in *Jennie Gerhardt,* as a norm by which to criticize the surrounding culture.

The same conjunction, used to similar ends, operates in Veblen's writings. "The only other instinctive factor of human nature that could with any likelihood dispute this primacy" of the instinct of workmanship, Veblen hypothesizes, is what he calls the "parental bent" (*IOW,* 25). As he puts it, "this ubiquitous parental instinct tends constantly to place motherhood in the foreground in all that concerns the common good, in as much as all that is worth while, humanly speaking, has its beginning here" (*IOW,* 94). Workmanship and the parental bent are intimately related; indeed, "[t]hey spend themselves on much the same concrete objective ends" so that Veblen finds it "a matter of extreme difficulty to draw a line between them" (*IOW,* 25). In fact, "the instinct of workmanship is in the main a propensity to work out the ends which the parental bent makes worth while" (*IOW,* 48). The reason for their close alignment is the same reason Veblen so values them: these two instincts reign "[c]hief among those instinctive dispositions that conduce directly to the material well-being of the race" (*IOW,* 25).

Gross oversimplifications of Veblen result from ignoring these positive instincts and focusing on his praise for technology; Veblen is as much humanist as technocrat.

Veblen's praise for workmanship and the parental bent follows the same logic as his condemnation of self-interest as the root of all evil. He considers self-interest to cause invidious behaviors and to define business enterprise. Furthermore, one of the main problems Veblen locates in neoclassic economic theory is its glorification of self-interest. But workmanship and the parental bent—in particular, the *maternal* bent—foster an outlook that is other-directed.[8]

Moreover, "parental solicitude in mankind has a much wider bearing" than reproduction and care for one's own offspring (*IOW,* 26); it looks out for the entire "community's future welfare" (*IOW,* 26, 27). Thus the parental bent, which fosters altruistic behavior and an ecological orientation, grounds crucial Veblenian value judgments. For instance, "it is a despicably inhuman thing for the current generation wilfully to make the way of life harder for the next generation, whether through neglect of due provision for their subsistence and proper training or through wasting their heritage of resources and opportunity by improvident greed and indolence" (*IOW,* 26). One could hardly ask a cultural critic to declare his values more forthrightly. Along with idle curiosity, the instinct of workmanship and parental bent occupy privileged positions in Veblen's thinking: they provide a locus of enduring positive value, and a vantage point from which to critique the pecuniary culture.

What Veblen defines as workmanship and the parental bent are the dominant characteristics of Jennie Gerhardt, as well as the reasons for Dreiser's reverence for her character—far more so than for her sexual availability, as some critics have charged.[9] Like Veblen, Dreiser associates work with the parental bent. The conjunction, evident in

---

8. In *Thorstein Veblen,* Riesman describes Veblen's thought originating in an "internalized colloquy between his parents: between one who calls for a hard, matter-of-fact, 'Darwinian,' appraisal of all phenomena and one who espouses the womanly qualities of peaceableness, uncompetitiveness, regard for the weak" (6-7). The neatness of the claim is appealing, but the loosely psychoanalytic frame insufficiently developed, and Riesman offers scant indication of its relevance to Veblen's life or work.

9. See, for instance, Hapke, *Tales,* 82; Richard Lehan, *Theodore Dreiser: His World and His Novels,* 87. On Jennie as Dreiser's maternal fantasy, see Susan Wolstenholme, "Brother Theodore, Hell on Women," 248. The most extensive critique of Dreiser's representation of women occurs in Gammel's *Sexualizing Power,* which adopts a Foucaultian framework to argue that Dreiser normalizes female sexuality and reaffirms conventional views of women.

Dreiser's descriptions of his mother, informs his portrayal of Jennie Gerhardt. The first time we see her, she and her mother are seeking employment. Even as a girl, Jennie is Mrs. Gerhardt's "right hand," an image that suggests her devotion both to her mother and to work (*JG*, 17). Jennie provides Mrs. Gerhardt with her "one stay," in contrast to Carrie Meeber, whose "one stay" is "pleasure" (*JG*, 149; *SC*, 34). Dreiser also insists on Jennie's own maternal nature. An "ideal mother" with "supreme motherly instincts," as the narrator calls her several times (*JG*, 97; cf. 108), Jennie is so fertile as to become pregnant after her first sexual encounter, and she gives birth effortlessly. An "indissoluble bond of affection" joins Jennie to her illegitimate daughter, Vesta, as to her own mother (*JG*, 378). The enjoyment Jennie takes in serving others, her natural affinity for her mother and child, the parental role she assumes for her father and dying lover, all cast her in a classically, even stereotypically, maternal role.

Jennie is the only major Dreiser character whose child figures both significantly and positively in her life. One scarcely thinks of Eugene Witla or Frank Cowperwood as good fathers. Cowperwood, as might be expected, "liked . . . the idea of self-duplication. It was almost acquisitive" (*F*, 57). He and his pale shadow, Witla, are more likely to seek girls as sexual conquests than to nurture them. For all her sexual experience, Carrie conspicuously lacks children. The specter of parenthood is clearly a curse to Roberta Alden and Clyde Griffiths. Jennie, however, finds it "a wonderful thing to be a mother—even when the family was shunned" (*JG*, 95–6), and it is largely Jennie's status as an unwed mother that causes her family to be shunned, a fact that Dreiser excoriates. The cultural criticism is clear, but it is nevertheless startling to come upon his association of a twentieth-century working woman with the "[c]ertain process of the All-mother, the great artificing wisdom of the power that in silence and darkness works and weaves" (*JG*, 92).

Jennie's maternal nature does not just make her a good mother (and daughter); she also manifests the altruism that Veblen associates with the parental bent. She responds to a beggar whom Lester doesn't even notice—reminiscent of a similar scene with Carrie and Hurstwood in *Sister Carrie*—with sympathy and generosity. "[Q]uick to see ragged clothes, worn shoes, care-lined faces" in others, Jennie instinctively gives to them (*JG*, 195). As Veblen remarks, "In persons highly gifted in [the parental bent] the impulse asserts itself to succour the helpless with one's own hands" (*IOW*, 33). The fate that Dreiser reserves for Jennie emphasizes her parental, altruistic, and workmanlike tendencies. "She also thought that some charitable organization might employ her or accept her services"—typically, she thinks of laboring for others, not of the recompense she might receive—"but she did not understand the new theory of charity which was then coming into general acceptance and practice—namely, only to help others to help themselves." Rather than so limited a notion of charity, she simply "believed in giving." Rather than follow "investment or . . . the devious ways of trade," Jennie enjoys "[t]he care of flowers, the care of children, the looking after and maintaining the order of a home" (*JG,* 397). She apprehends, in short, the connections among human beings and other living things. Jennie demonstrates what today appears as a nascent ecological consciousness.

Jennie is equally unusual in the Dreiser canon for her devotion to working. Most of Dreiser's characters dislike any sort of manual labor; in fact, they find it odious. Although Carrie's "sympathies were ever with that underworld of toil from which she had so recently sprung," her aversion to the work available to her in Chicago propels the novel forward (*SC,* 140). The Pennsylvania edition is even more explicit: Carrie finds work "absolutely nauseating" (Penn *SC,* 39). For Carrie, the only thing worse than doing manual labor is thinking about it: "Toil, now that she was free of it, seemed even a more desolate thing than when she was part of it" (*SC,* 140). Hurstwood's resistance to work after he loses his ceremonial position as a saloon manager is equally instrumental to the plot of *Sister Carrie. An American Tragedy* is also saturated with aversive feelings toward work. Crucial to Clyde's characterization is the narrator's early remark that "true to the standard of the American youth, or the general American attitude toward life, [Clyde] felt himself above the type of labor which was purely manual" (*AAT,* 18). Similar to Carrie and Clyde, Roberta considers work at the hosiery mill "beneath her" (*AAT,* 245). And Clyde is initially attracted to Roberta because she appears to be "really above the type of work she was seeking" (*AAT,* 241).

Dreiser's biography again illuminates his characters' predilections. In *An Amateur Laborer,* the unfinished account of his breakdown after the "suppression" of *Sister Carrie,* the onset of his neurasthenia, and the death of his father, Dreiser records his own deeply conflicted feelings about manual labor. (Dreiser also fictionalizes this episode of his life in *The "Genius"* and sketches it in *Twelve Men.*) Securing work as a day laborer for the New York Central Railroad, he notes the "weariness and regret" in others' faces the first day on the job (*AL,* 110). He betrays his ambivalence about taking the job when he calls it an experiment in the "recreative career of labor" (*AL,* 113). Contradictory emotions surface quickly, as Dreiser wonders, "Was it that the mere sight of work was so offensive to me? No, I thought; rather it is beautiful, but there is something there, uneasiness or dissatisfaction or the old human ache that makes one wish to be somewhere else" (*AL,* 105-6). Similarly, he says, "The idea of more work of this kind—hourly work—daily work—yearly work, had an appalling effect on me," but in the following paragraph he voices a competing sentiment: "And yet in the midst of these very reflections there was something infinitely sweet. . . . Here was an illustration of the meaning of life. . . . Man was made to work. He was made to fit in with the great scheme of labor and nothing but labor was satisfactory to it" (*AL,* 116).[10]

Dreiser's recollection of manual labor echoes the description of his mother "stitching, stitching, stitching," except that in *An Amateur Laborer,* as in most of his fiction, work takes on an unpleasant cast. Dreiser's ambivalence suggests a conflict between the need of the human animal to do useful work and a ceremonial aversion to harming one's self-esteem by performing degrading tasks. Veblen states this conflict explicitly: "Manual labour . . . is of course on a precarious footing as regards respectability" (*TLC,* 232). Even Jennie is embarrassed over her need to ask if she can clean the Columbus hotel—"not because it irritated her to work, but because she hated people to know" (*JG,* 3).

The antipathy toward manual labor expressed by Dreiser and so many of his characters, as well as the reason Jennie doesn't want people to know she must perform menial tasks, illustrate what Veblen calls the irksomeness of labor. Although workmanship is one of the few positive characteristics Veblen locates in human nature, he also must account for why modern humans tend to find labor so distasteful. What makes manual labor "dreary" and "inexpressibly wearisome," to use Dreiser's words from *An Amateur Laborer* (160), is the irksomeness of labor.

This distinction between workmanship and labor is, according to Hannah Arendt, an aspect of the human condition. She notes that

---

10. See also *Dreiser Looks at Russia,* in which he praises the Soviet leaders' realization "that the best thing for everybody is work in some form . . . and the worst thing idleness" (11; cf. *DLR,* 14-15 on the pleasure of work).

European languages have different words for labor and work. But Veblen, in contrast to orthodox economists (as well as contra Freud, who asserted a "natural human aversion to work"), declares that the "aversion to labor is in great part a conventional aversion only." (*ECO*, 81).[11] The crucial word is *conventional*. In an early essay that is one of his richest, "The Instinct of Workmanship and the Irksomeness of Labor" (1898), Veblen takes on "the commonplac[e] of the received economic theory that work is irksome," which fosters the idea that "men desire above all things to get the goods produced by labor and to avoid the labor by which the goods are produced." He declares that idea impossible on evolutionary grounds, for "there is no chance for the survival of a species gifted with such an aversion to the furtherance of its own life process" (*ECO*, 78–79). Rather than being innate, instinctual, or natural, Veblen insists the human aversion to labor (the perception of its irksomeness) is a response learned relatively late in human history—thus its "conventionality." The irksomeness of labor, an institutional excrescence, deforms the instinctive human desire to work.

Veblen's distinction between instinct and convention follows William James's contention that "habit" can modify "instinct." Thus James's famous declaration that habit is "the enormous fly-wheel of society, its most precious conservative agent."[12] To establish that the irksomeness of labor is conventional, not instinctual, Veblen slyly points out that destructive and ceremonial exertions (he mentions warfare and sports) are highly regarded (*ECO*, 95). He elsewhere labels the two spheres of work: destructive "exploit" is esteemed, whereas productive "drudgery" is looked down upon (see, for instance, *TLC*, 13). The ceremonial/substantive dichotomy recalls, again, business versus industry.

As is typical of Veblen, he explains the irksomeness of labor by using "scientific" reasoning to back up value judgments. Again, his institutional viewpoint allows him to embrace ideas from many disciplines. As C. E. Ayres notes, Veblen "was virtually unique among economists [of his time] for his acquaintance with anthropological literature and his assimilation of such anthropological concepts as that of culture." Veblen gives not only chronological but also axiological precedence to the instinct of workmanship and closely related parental bent over the

---

11. Hannah Arendt, *The Human Condition*, 80; Sigmund Freud quoted in David D. Gilmore, *Manhood in the Making: Cultural Conceptions of Masculinity*, 98.

12. James, *Principles* 2:395, 402; *Principles* 1:121.

otiose, invidious and pecuniary behaviors that develop later. As Veblen puts it, the "indispensable native traits of the successful [human] race [are] the parental bent and the sense of workmanship, rather than those instinctive traits that make for disturbance of the peace" (*IOW,* 122). David Riesman suggests the way to make sense of this problematic point in Veblen's thinking, where some instincts seem to be more essentially human than others, is to see predatory emulation only as a "quasi-instinct," and consider that Veblen believed there was a " 'natural' human nature . . . undistorted by emulation."[13]

In this respect Veblen goes further than Marx in making the concept of work a central locus of value. While Marx's labor theory of value gives work preeminence in the construction of economic value, Veblen considers cultural and psychological attitudes toward work, even more than class position, to be defining factors in human behavior.[14] As Veblen distinguishes his position from Marx's, "It is a question not so much of possessions as of employments; not of relative wealth, but of work. It is a question of work because it is a question of habits of thought, and work shapes the habits of thought" (*TBE,* 348). Thus Veblen's economic materialism gives work the power to determine consciousness, where Marx would look to social class. Indeed, remarks Veblen, "[t]his interest in work differentiates the workman from the

---

13. C. E. Ayres, "Veblen's Theory of Instincts Reconsidered," 25; Riesman, *Thorstein Veblen,* 64, 156.

14. Veblen disputes the labor theory of value in a note to "The Instinct of Workmanship and the Irksomeness of Labor"; see 145–46 n 2. On Veblen's critique of Marx's labor theory of value, see Arthur Vidich, "Veblen and the PostKeynesian Political Economy."

Veblen distances himself from socialism when he argues that classes divide "not between those who have something and those who have nothing—as many socialists would be inclined to describe it—but between those who own wealth enough to make it count, and those who do not" (*VI,* 160-61). Veblen locates the real division "between those who live on free income and those who live by work. . . . It is sometimes spoken of in this bearing—particularly by certain socialists—as a division between those who do no useful work and those who do; but this would be a hasty generalisation, since not a few of those persons who have no assured free income also do no work that is of material use, as e.g., menial servants" (*VI,* 161).

But Veblen and Marx come close to agreement in seeing modern people as tragically alienated from their human nature. As Marx puts it in "Critical Marginal Notes on the Article 'The King of Prussia and Social Reform,' " "*Human nature* is the *true community* of men. The disastrous isolation from this essential nature is incomparably more universal, more intolerable, more dreadful, and more contradictory, than isolation from the political community" (131). According to Marx, because "man is estranged from the product of his labour," the result is "the *estrangement of man* from *man*" ("Economic and Philosophic Manuscripts of 1844," 77).

criminal on the one hand, and from the captain of industry on the other" (*TLC,* 242)—from, we might say, Clyde or Hurstwood on one hand, Cowperwood on the other. Jennie Gerhardt represents the excluded middle that both Veblen and Dreiser admire.

The predatory status system that Veblen loathes accounts, once again, for the unwholesome and imbecilic traits of modern humans. The "addiction to work becomes a mark of inferiority and therefore discreditable" only *after* "work becomes distasteful to all men instructed in the proprieties of the pecuniary culture" (*IOW,* 174). Veblen resorts to a biblical allusion to emphasize the point: "If such an aversion to useful effort is an integral part of human nature, then the trail of the Edenic serpent should be plain to all men" (*ECO,* 78). Veblen's invocation of Christian typology here is striking in its absence of irony, but as T. W. Adorno remarks, Veblen is a "puritan *malgré lui-même,*" particularly for his glorification of workmanship.[15]

Because the concept of instinct is instrumental in upholding Veblen's values, he doggedly held on to it even after the idea began falling out of favor in the scientific community. He even uses the troublesome word in the title of one of his books, *The Instinct of Workmanship and the State of the Industrial Arts.* Veblen insinuates that aversion to the idea of instinct is a form of pedantry: "to dispense with it [the concept of instinct] . . . is an untoward move in that it deprives the student of the free use of this familiar term in its familiar sense and therefore constrains him to bring the indispensable concept of instinct in again surreptitiously under cover of some unfamiliar term or some terminological circumlocution" (*IOW,* 28 n). Instincts such as workmanship and the parental bent constitute what Veblen warily describes as " 'human nature' " or " 'spiritual nature,' " his quotation marks indicating his reluctance to endorse such loaded terms (*IOW,* 14). Furthermore, "all instinctive action is intelligent" and "teleological" in that it "involves holding to a purpose" (*IOW,* 31).

Veblen was well aware of the change of tides. He insists that while the concept of instinct may seem "lax and shifty" to biologists and imprecise to psychologists, it enables his own "inquiry into the nature and causes of the growth of institutions" (*IOW,* 1–2). Furthermore, reasons Veblen, "instinct" retains a powerful and unambiguous commonsensical meaning even if academics pretend not to understand it.

---

15. Adorno, "Veblen's Attack," 83.

In 1935 C. E. Ayres wrote to Joseph Dorfman that Veblen "definitely" did not believe literally in instinct psychology. Ayres recalls Veblen's asking if he had ever noticed how he defined "instinct." When Ayres said no, Veblen "remarked that no such definition appears [in his work] because if you define instincts exactly, 'there ain't no such animal.'"[16] As Ayres explains Veblen's use of instinct theory, the scientific validity of the concept matters less than the antithesis Veblen constructs from it: humans have contradictory drives toward workmanship and ceremonialism, whatever one chooses to call them.[17]

The basic complaint leveled against instinct was (and has generally remained since Veblen's day) that the biological emphasis obscures the important role played by culture in human behavior.[18] The reification of biology over culture, particularly when used to formulate social policy, can indeed lead to pernicious results. But those who have criticized Veblen's deployment of instinct as naive biologism miss the point, for he uses the concept strategically to advance his decidedly cultural criticism. The problems he locates in human behavior are unquestionably cultural—caused by the "pecuniary culture" and "institutional exigencies" (*IOW,* 171, 172).[19] By elevating the instinct of workmanship over the irksomeness of labor in terms both evolutionary and axiological, Veblen reverses the values he finds upheld in American society. His use of the concept of instinct to glorify workmanship and the parental bent also suggests that he saw humanity in more favorable terms than most commentators have realized.

16. C. E. Ayres to Joseph Dorfman, Joseph Dorfman papers, Thorstein Veblen Correspondence, Columbia University.

17. C. E. Ayres, "Veblen's Theory of Instincts Reconsidered." Veblen did not attach a rigid meaning to "instinct" such as the word generally connotes. He assumes an "endless complication and contamination of instinctive elements" (*IOW,* 29; on the "contamination" of instincts see especially *IOW,* chap. 2). While more fundamental than invidious and predatory tendencies, the instinct of workmanship is, according to Veblen, easily obstructed (*IOW,* 51).

18. For a recent description of this view, see Degler, *In Search of Human Nature,* 156-62.

19. "The Mutation Theory and the Blond Race" shows Veblen's nuanced understanding of the intersection of biology and culture. He notes that discussions of (supposedly biological) race were generally already involved with cultural considerations, although commentators rarely acknowledged that fact (*POS,* 458). He argues that culture can change race (*POS,* 462) and notes that so-called "national and local types" that seem to have "acquired a degree of permanence" in fact only "simulate racial characters" and are matters of "habit" and thus "institutional element[s] rather than . . . characteristics of race" (*POS,* 472-73).

Divisions of Labor, Divisions of Gender

Although the conventional irksomeness of labor is exemplified by one of Veblen's best known constructs—the woman who conspicuously consumes goods to advertise her husband's pecuniary success, as immortalized in *The Theory of the Leisure Class* (1899)—he believed that women inherently had a greater portion of workmanship than men. According to Veblen, women occupied "the chief place in the technological scheme" during earliest human culture (*IOW,* 94). *The Instinct of Workmanship* contains many parenthetical comments to set the record straight. For instance, "Through long ages of work and play men (perhaps primarily women) learned the difficult and delicate crafts of husbandry" (69); "The man (more often perhaps the woman) who busies himself with the beginnings of plant and animal breeding. . . ." (76); "No one is competent to acquire such mastery of all the lines of industry included in the general scheme as would enable him (or her) to transmit the state of the industrial arts to succeeding generations unimpaired at all points" (107); "The workman—more typically perhaps the workwoman— . . . is a 'productive agent' " (144); the early, "savage" social order was "peaceable, non-coercive . . . , with maternal descent and mother-goddesses, and without much property rights, accumulated wealth or pecuniary distinction of class" (153). Unlike the followers of Auguste Comte who posited an unbroken chain of progress from savagery through barbarism to civilization, Veblen depicts the matrifocal savage state as the golden age.

Veblen is no more exact in his dating than contemporary evolutionary sociologists such as Lester Ward, but at one point he cites the early Neolithic as the era of transition from peaceable savagery to predatory barbarism (*IOW,* 149). Disputing the Hobbesian view that the earliest human communities existed in a state of war and continual fear (*IOW,* 123), Veblen contends that the "savage mode of life" involves "considerable group solidarity . . . , living very near the soil, and [being] unremittingly dependent for their daily life on the workmanlike efficiency of all the members of the group." Most important for survival is the "propensity unselfishly and impersonally to make the most of the material means at hand" (*IOW,* 36). Veblen's anthropological assumptions led David W. Noble to remark on the "paradox of progress and primitivism" in Veblen: for a theorist so enamored of technology, Veblen's veneration of a "savage" golden age is indeed remarkable. Noble sees Veblen's views on the savage social order as revealing his "hidden belief in a normal man, a man above and outside historical change."[20]

But no less a historical mind than Darwin's licenses Veblen's chronology of the development of instincts. In *The Descent of Man* (1871) Darwin hypothesizes that the "social and maternal instincts" are the most enduring and persistent, whereas the self-regarding instincts emerge later in human evolution. Darwin's ranking of instincts also betrays definite preferences: the "lower" instincts "relate chiefly to self" and "public opinion" while the "higher" "social instincts . . . relate to the welfare of others." Likewise, according to Veblen, during the early stage characterized by communal interdependence and peace, the two human "instincts which make directly for the material welfare of the community"—the instinct of workmanship and parental bent—were allowed free rein (*IOW,* 25).[21]

Dreiser draws a similarly prelapsarian portrait in *Jennie Gerhardt*.[22] The title character resembles one of Veblen's peaceable savages living in a world of predatory barbarians. That tension between Jennie and the surrounding culture informs the entire novel, accounting both for Dreiser's often sentimental praise of Jennie and for his criticism of her culture. The coexistence of sentimentalism with criticism becomes clear in the following passage:

> The spirit of Jennie—who shall express it? This daughter of poverty . . . was a creature of a mellowness. . . . There are natures. . . . [which] see a conformable and perfect world. Trees, flowers, the world of sound and the world of color. These are the valued inheritance of their state. If no one said to them "Mine," they would wander radiantly forth, singing the song which all the earth may some day hope to hear. It is the song of goodness.
> Caged in the world of the material, however, such a nature is almost invariably an anomaly. (*JG,* 16)

As Dreiser depicts Jennie, she apprehends a unified and coherent world not yet truncated by divisive claims of "Mine." Ownership, Veblen

---

20. Noble, "Dreiser and Veblen," 144–46.

21. Darwin, *The Descent of Man,* chap. 3 (87, 100). Herbert Spencer, in contrast, believes humans evolved from disordered clusters of self-sufficient individuals to ordered groups that assume hierarchies. Spencer, in short, differs from Darwin and Veblen in seeing the "social" tendencies emerging relatively late in human evolution. See *First Principles,* 342–43.

22. The prelapsarian quality of *Jennie Gerhardt* has been noted by previous commentators, in particular Wadlington, "Pathos and Dreiser."

says, "is self-regarding, of course, and the rights of ownership are of a personal, invidious, differential, emulative nature" (*IOW,* 172). And so Jennie, innocent of these motives, is tragically underrated by the barbarians around her.[23]

Veblen's anthropological assumptions also account for why women's domestic work, such as Jennie continuously performs, falls into particular disrepute. According to Veblen, the decline into barbarism marks the fall of women's status as producers. During the barbaric era, male ceremonial exploits (fighting battles, competing in contests, capturing women, and so on), which benefits only specific individuals, acquires prestige, while women's work, which benefits the group, is redefined as "drudgery" and loses status (*TLC,* 10–15; *ECO,* 94). The irksomeness of labor reflects the association of drudgery with "inferiority," weakness— and with women (*IOW,* 174; *TLC,* 36). As Veblen says, the "marked distinction . . . between the occupations of men and women . . . is of an invidious character"; indeed, he implies that the construction of gender differences into hierarchical rankings forms the root of all invidious behavior (*TLC,* 4).

For Dreiser, also, the irksomeness associated especially with women's work is fraught with status claims. He presses for acknowledgment of the low-status work performed by Jennie's mother: "Every day Mrs. Gerhardt, who worked like a servant and received absolutely no compensation either in clothes, amusements or anything else, arose in the morning while the others slept, and built the fire" (*JG,* 108). His sympathetic treatment echoes Charlotte Perkins Gilman's argument in *Women and Economics* (1898) that the mother is the "worker *par excellence,* but her work is not such as to affect her economic status."[24] Dreiser's contempt for the way women's serviceable work is looked down upon becomes particularly pointed in the scene where Lester's

---

23. In *Homage,* Warren notes that "the Senator and Lester, unlike Drouet and Hurstwood, who are totally fulfilled in the world of success, repudiate its values, and find values in Jennie that the practical world scorns" (44). My interpretation accords with both John B. Humma's "*Jennie Gerhardt* and the Dream of the Pastoral" and Christopher P. Wilson's "Labor and Capital in *Jennie Gerhardt.*" Humma argues that Jennie, who represents the pastoral ethic, is "the one major character in Theodore Dreiser's fiction whom Dreiser without qualification and with perfect sincerity approves" (157), while Wilson locates *Jennie Gerhardt*'s positive values in the "older, artisan-based universe" represented by the Gerhardts, which Dreiser uses to deliver a "moral critique" of the Kane family as an embodiment of the modern "corporate sector" (107, 111, 103).

24. Gilman, *Women and Economics,* 21.

sister, Louise, discovers his ménage with Jennie. Louise expresses her disgust by invoking both her brother's social position and the irksomeness of labor: " 'I should think . . . that you of all men would be above anything like this—and that with a woman so obviously beneath you. Why I thought she was—' she was again going to add 'your housekeeper' " (*JG*, 227). Louise's unspoken words demonstrate Veblen's claim that members of "the better class" "sense . . . ceremonial uncleanness attaching . . . to the occupations which are associated . . . with menial service" (*TLC*, 37).

Ever since productive work was defined as irksome, menial service has been considered inappropriate—indeed, unfeminine—for women with status claims. One does not expect to find the lady on Millionaires' Row covered in soapsuds and sweat. As Veblen voices the familiar sentiment, "It grates painfully on our nerves to contemplate the necessity of any well-bred woman's earning a livelihood by useful work. It is not 'woman's sphere' " (*TLC*, 179). Again, "The good and beautiful scheme of life, then—that is to say the scheme to which we are habituated—assigns to the woman a 'sphere' ancillary to the activity of the man; and it is felt that any departure from the traditions of her assigned round of duties is unwomanly" (*TLC*, 354). As he does with economic constructs he considers dubious, such as "ordinary profits," Veblen uses quotation marks here to draw attention to the ceremonial nature of "women's sphere."

Lester endorses this class-based notion of women's sphere when he tries to restrain Jennie from doing her own housework: "She would have done most things herself, had Lester not repeatedly cautioned her not to. 'There's just one way to do this thing,' he insisted. 'Get someone else to do it' " (*JG*, 268). Extolling her to adopt a managerial role and delegate her work out, Lester would heighten Jennie's status at the cost of her serviceability. Interestingly, it is only under the influence of Lester that Jennie stops working so much, and even then, leisure seems abnormal to her. (Carrie's situation differs in that she detests the work available to her in Chicago; nevertheless, she, like Jennie, also stops working only under the influence of a man.[25]) In particular, Jennie considers the cook a "useless extravagance" (*JG*, 267). Veblen

---

25. The classic literary reference point from this period of an industrious woman chained to an idle man is Trina Sieppe McTeague in Norris's *McTeague*. Appropriately, the ex-dentist murders his wife at her place of work.

would not be surprised at her opposition to extravagance, for he sees the "common-sense" resistance to waste as "itself an outcropping of the instinct of workmanship" (*TLC*, 98).

The "spirit of Jennie" passage, which directly precedes the "Caged in the world of the material" paragraph, concludes: "It is of such that the bondservants are made" (*JG*, 16). Her prelapsarian spirit is enthralled by the pecuniary culture. The domestic service central to Jennie's character operates as an important cultural marker to Veblen. His essay on "Menial Servants during the Period of the War" (1918)—an outrageous argument delivered in deadpan which bears comparison with Jonathan Swift's "A Modest Proposal"—advances the idea that because of the labor shortage during World War I, the government should tax employers of domestic servants at the rate of 100 percent of wages for the first, 200 percent for the second, and so on. (One might expect Veblen also to suggest that employers would welcome this opportunity for the display of conspicuous waste.) The essay pivots on Veblen's explanation of why people hire servants in the first place: because " 'Servant' implies 'Master,' of course" (*ECO*, 270). Similarly, in *Leisure Class* he remarks, "The pervading norm in the predatory community's scheme of life is the relation of superior and inferior, noble and base, dominant and subservient . . . , master and slave" (*TLC*, 301).

The appeal of establishing his mastery by having a servant around is one of several motives attracting Jennie's first lover to her. Senator Brander's position at the hotel where he first encounters Jennie places him well above her: his appearance "marked him at once . . . as some one of importance" (*JG*, 8). But the invidious contrast with her poverty only heightens his sense of self-importance: "The girl's poor clothes and her wondering admiration for his state affected him. He felt again that thing which she had made him feel before—the far way he had come along the path of comfort. How high up he was in the world, indeed!" (*JG*, 22). Shortly, Brander feels again "that same sensation . . . the far cry between her estate and his. It was something to be a senator tonight, here where these children were picking coal" (*JG*, 29). Veblen explains the pleasurable sensation such as Brander feels as the result of invidious comparisons: "the weakness of one party (in the pecuniary respect) is as much to the point as the strength of the other,—the two being substantially the same fact" (*IOW*, 191). Jennie's poverty enhances Brander's sense of wealth and power.

Veblen notes also that "[a]n aristocratic (or servile) scheme of life must necessarily run in invidious terms, since that is the whole meaning of the phenomenon" (*IOW,* 183). When Brander offers Jennie cash, he is motivated by contradictory feelings of lust, altruism, and the desire to see the flattering image of his mastery which can only be reflected in a servant's eyes. Buying her would secure Brander's invidious superiority, and that is just what he seeks to do: " 'And here,'—he reached for his purse and took from it a hundred dollars. . . . 'You're my girl now—remember that. You belong to me' " (*JG,* 74–75). Veblen considers it "sufficiently plain, to any one who cares to see, that our bearing towards menials and other pecuniarily dependent inferiors is the bearing of the superior member in a relation of status, though its manifestation is often greatly modified and softened from the original expression of crude dominance" (*TLC,* 52). The master/slave dichotomy epitomizes the invidious method of self-construction.

The association of Jennie's passivity with her sex appeal, suggested first in the episode with Brander and many times thereafter, makes it easy to overlook the more progressive dimensions of Dreiser's analysis of heterosexuality. Dreiser's treatment of erotic attraction in *Jennie Gerhardt* focuses sharply, in Veblenian fashion, on what is in effect a class difference between men and women. Dreiser's attention to Brander's maintenance of his self-esteem through Jennie's favorable notice anticipates the famous passage in *A Room of One's Own* (1929) where Virginia Woolf describes men constructing favorable self-images by the reflected gaze of the women who love them: "Women have served all these centuries as looking-glasses possessing the magic and delicious power of reflecting the figure of man at twice its natural size."[26] So does Brander feel "exceedingly young sometimes as he talked to this girl [Jennie]" (*JG,* 24). Brander's sense of rejuvenation is of course the paradigmatic sexual response for a Dreiser male (character or author). The predatory dimensions of such a relationship are obvious; what has been overlooked is Dreiser's sensitivity to how dependent men are on women, even (or perhaps especially) on young women.

---

26. Virginia Woolf, *A Room of One's Own,* 35. Dreiser's and Veblen's analysis of power as produced by class and gender difference also anticipates Foucault's insistence in *The History of Sexuality* that "relations of power are not in a position of exteriority with respect to other types of relationships (economic processes, knowledge relationships, sexual relations), but are immanent in the latter; they are the immediate effects of the divisions, inequalities, and disequilibruims which occur in the latter" (94).

During his brief appearance in the novel, Brander becomes associated with his walking stick, which he carries "evidently more for the pleasure of the thing than anything else" (*JG*, 8). Jennie notes that the cane has a gold head, and the neighbor who tells old Gerhardt of his daughter's indiscretions also mentions the walking stick (8, 54). The conspicuous phallic symbolism (and a gold penis, no less) reiterates the connection between aggressive masculinity and invidious class distinction. Veblen analyzes the walking stick along similar lines:

> The walking-stick serves the purpose of an advertisement that the bearer's hands are employed otherwise than in useful effort, and it therefore has utility as an evidence of leisure. But it is also a weapon, and it meets a felt need of barbarian man on that ground. The handling of so tangible and primitive a means of offence is very comforting to any one who is gifted with even a moderate share of ferocity. (*TLC*, 265; see also *TLC*, 171)

Again a division of sexual function figures in to Veblen's reasoning: women, he claims, do not carry canes for ceremonial reasons, only when infirmity makes one necessary (*TLC*, 265).

Jennie's relationship with her second lover, named not Stick but Kane, follows the more familiar component of Veblen's analysis of gender roles during and beyond the barbaric era, with wives conspicuously consuming goods to advertise their husbands' wealth. Like Brander, Lester apprehends the invidious status distinction between himself and Jennie (returning to his home in Cincinnati, Lester notes "the distinctive nature of his home life, so different from the quality of the liaison he had fallen on in Cleveland" [*JG*, 141]), and again like his predecessor, Lester sees a pleasing image of his prowess reflected in Jennie's response (*JG*, 128). But Lester feels even more predatory and proprietary toward Jennie. When Lester appears in the novel his attitude toward women is decidedly materialistic; he equates courtship with shopping: " 'I want to browse around a little while yet,' " he responds to those who press him to settle down (*JG*, 122). According to Veblen, the sexual response of modern males always reveals the impulse for ownership. Lester, apparently, sees Jennie as a suitable trophy, for he quickly becomes eager to win her as a "prize" (*JG*, 135, 154). "Why shouldn't he try to seize her?" Lester asks himself (*JG* 124). Like Brander, he considers Jennie a possession to own: "You belong to me," he declares shortly after meeting her, and she soon comes to feel that she does (*JG*, 123, 124). Lester, "athletic, direct and vigorous," is even more the able barbarian than Brander (*JG*, 121). Although strongly attracted to this powerful man, Jennie also feels "horrified" by his distinctly predatory tendencies; she imagines herself "like a bird in the grasp of a cat" (*JG*, 123). By the end of the novel, however, Jennie's "fondest memories were of the days . . . when he had seized her, much as the cave man had seized his mate—by force" (*JG*, 409).

What Veblen describes as "seizure by prowess" and "the prescriptive tenure of whatever one has acquired" characterizes Lester and Jennie's relationship (*IOW*, 202). "[T]he women so held in constraint and in evidence will commonly fall into a conventionally recognised marriage relation with their captor," Veblen claims (*ECO*, 47). Lester encourages Jennie to adopt the role of vicarious, conspicuous, trophy wife. Besides asking her to depend on him economically instead of working (*JG*, 156–57), Lester initiates Jennie into conspicuousness. "[V]ery proud of his prize and anxious to make her look beautiful," Lester promises Jennie "to show you what you can be made to look like" (*JG*, 160). (Strictly speaking, she was first exposed to the regime of status as a servant at Mrs. Bracebridge's. Jennie's wealthy mistress explains that "Life is a battle" and under Mrs. Bracebridge's tutelage, Jennie develops a "faint perception of hierarchies and powers" (*JG*, 109). The conventionalization succeeds in part. When Jennie walks down the street as Lester's trophy, people turn to stare and comment, "That is a stunning woman that man has with him" (*JG*, 167). Thus, Lester manages "to make her look like someone truly worthy of him" (*JG*, 167).

But the trophy remains an inauthentic gentlewoman, the marriage ceremonial in Veblen's sense of make-believe display. Dreiser uses Jennie's make-believe status as wife not to condemn her but to expose the values of the surrounding society. Jennie so appeals to men because she is, as Lester formulates the paradox, such a "lady-like" "servant" (*JG*, 124). As readers of *Leisure Class* will recall, Veblen considers any wife living in a "household with a male head" to be "still primarily a servant"; indeed the "lady" of the rich man's house is but the "chief menial" (*TLC*, 60, 182). In *Jennie Gerhardt*, Dreiser makes the same point from a different perspective, for Jennie is far more gentle, and genteel, than the thoroughbred women and men who look down upon her. Thus Dreiser's second novel provides an important chapter in his lifelong concern with what he refers to in *The Titan* as "our built-up system of ethics relating to property in women" (121).[27]

Moral Bankruptcy and Pecuniary Masculinity

A Veblenian analysis can illuminate Lester's own considerable personal conflicts, not all of which have to do with Jennie. Dreiser so richly describes Lester's turmoil as to lead numerous critics of the 1911 edition to see him as the real center of the novel. As Donald Pizer observes, "Lester, the man of action and strength, is unable to make a decision," and on this Hamlet-like indecision much of the novel seems to rest.[28]

Lester's conflict emerges dramatically in his relationship with his brother. "[A]n uninvited standard of conduct thrust upon [Lester]," Robert functions as his brother's double, encouraging their family, other characters in the novel, and Dreiser, invidiously to compare the two (*JG,* 169). Dreiser uses the fraternal doubling to deepen his characterization of Lester's conflicts, anticipating the more complex doubling— between Clyde and Gilbert, and between Clyde and Roberta—of *An American Tragedy*. The opposition between the Kane brothers functions, as such contrasts always do in Veblen, to demonstrate the moral bankruptcy of the qualities leading to success in the eyes of the world. Because Lester becomes the victim of the same sort of status reckoning as Jennie does (though on a far lesser scale), he becomes a more understandable, even a more appealing, character.

The contrast is largely played out in the brothers' different approaches to business. Dreiser does not admire the conformist Robert Kane as he does the convention-defying Frank Cowperwood, whose

---

27. See also Dreiser's comparison of the social position of American and Soviet women (*DLR,* 169). In his introduction to Dreiser's *American Diaries,* Thomas P. Riggio suggests that women represented social change to Dreiser.

28. Pizer, *Novels,* 114. The Pennsylvania edition makes it difficult, if not impossible, to see Lester as the central character. In his preface to the Pennsylvania edition of *Jennie Gerhardt,* Riggio asks concerning the revisions for the 1911 text: "Why . . . did the character of Jennie undergo a more thorough revision than that of Lester? What assumptions about gender dominated the literary marketplace and led the Harpers editors to turn Jennie into the more passive, nebulously drawn figure upon which much modern criticism has focused?" (x–xi).

intransigence Dreiser identifies as a form of cultural criticism. Robert is more purely the narrow businessman deplored by Veblen than is Cowperwood. According to Veblen, "the spirit of American business is a spirit of quietism, caution, compromise, collusion, and chicane" (*HLA*, 70-71), a spirit that Robert embodies. For instance, as he plans horizontal integration to expand his father's carriage business, Robert figures that "if he could buy secretly into the stock of several othe[r companies], he could exercise a powerful influence toward the general combination which he hoped to effect. . . . He did not at all object to waiting" (*JG*, 189). "Watchful waiting" is another characteristic that Veblen, in one of his most savage descriptions, attributes to businessmen:

> Doubtless this form of words, "watchful waiting" [which constitutes the wisdom of business enterprise], will have been employed in the first instance to describe the frame of mind of a toad who has reached years of discretion and has found his appointed place among some frequented run where many flies and spiders pass and repass on their way . . . ; but by an easy turn of speech it has also been found suitable to describe the sane strategy of that mature order of captains of industry. . . . There is a certain bland sufficiency spread across the face of such a toad so circumstanced, while his comely personal bulk gives assurance of a pyramidal stability of principles. (*AO*, 109-10 n)

Dreiser chooses a reptile instead of an amphibian to describe Robert, who knows that being "snaky" is the key to business success (*JG*, 303). His "chill, persistent chase of the almighty dollar" makes Robert a master at what Veblen calls pecuniary accountancy (*JG*, 170). Robert's "Asiatic perception of the main chance" (*JG*, 169) links him explicitly with Veblen's analysis of business on two counts: the phrase "main chance" is a distinctive Veblenism, and Veblen prophetically analyzed the commercial prowess of the Japanese.[29]

Lester's "secret contempt" for Robert's business principles makes him a far more likeable character, but this contempt (and the equivocation it engenders) also disables him socially and financially, and the consequences further Dreiser's cultural criticism. Despite what Jennie believes to be Lester's opposition to children he demonstrates that he has some of the parental bent in his debate with Robert over old employees. Lester urges a paternalistic, "humane course" while

---

29. See *Imperial Germany*, 201; and "The Opportunity of Japan" (reprinted in *ECO*).

his brother wants to clear out the "dead wood" of the company (*JG,* 170). Living with Jennie further humanizes Lester; her benevolent philosophy "appealed to him as a big, decent way to take life, even if it did eliminate aggressiveness and the ability to gather material things" (*JG,* 195). But such kindly motives, Veblen ironically notes, "detract from business efficiency, and an undue yielding to them on the part of business men is to be deprecated as an infirmity" (*TBE,* 41). After the financial and social pressures on Lester grow to be almost intolerable, the narrator explains,

> The trouble with Lester was that, while blessed with a fine imagination and considerable insight, he lacked that ruthless, narrow-minded insistence on his individual superiority which is the necessary element in almost every great business success. To be a forceful figure in the business world means, as a rule, to be an individual of one idea largely, and that idea the God-given one that life has destined you for a tremendous future. (*JG,* 305)

Dreiser's irony is quietly effective: the "trouble" is Lester's "lacking" ruthlessness, megalomania, and tunnel vision. In contrast, Robert's one-dimensional criterion—"The business of this concern is to make money" (*JG,* 170)—helps him to succeed.

Dreiser uses the brothers' differing business principles to explore the depths of Lester's psyche. Even early in the novel, Lester knows that "Robert was obviously beating [him] in the game of life" (*JG,* 186), and his sense of self-worth becomes nearly as confused as Clyde Griffiths's does. As with Clyde, Lester's self-image is largely a matter of how he looks, or more precisely, how he sees himself reflected in others' eyes. The invidious comparison with his brother causes Lester great anxiety: "Although Lester did not consider himself either a mental, moral or any other kind of failure at this time, nevertheless this great illusive glitter of the other man was in his eye" (*JG,* 187).

This "glitter of the other man," so like the Clyde-Gilbert dynamic in *An American Tragedy,* suggests Dreiser's sense of the precariousness of male identity. Gender anxieties afflict Dreiser's male characters far more pointedly than his females. Insofar as Dreiser felt his own father to be a failure as a man, and was notoriously eager to demonstrate his own sexual prowess, this concern over masculinity comes as no surprise. The evidence of the novels is telling: Hurstwood fails at the male provider role after washing out in the manly realm of business;

Clyde, a momma's boy, has to overcome his initial fear of women, only to discover through Roberta's pregnancy the awful power that women seem to have over men; most revealingly, Eugene Witla faces the disconcerting prospect that sex with his wife drains him of his artistic talent. Dreiser's female characters may be mistreated by father and lovers (Jennie), exploited by capitalism (Roberta, the early Carrie), fetishized (Sondra, Berenice Fleming, Suzanne Dale), but their femininity is never in question.

Veblen also addresses male gender anxiety. He reasons that the earliest form of male honor depended on invidious gender distinctions—which could not, after all, be formulated without women—and that contemporary male status is equally dependent on women's advertising their husbands' prowess. The real reason for women's leading "vicarious" (*TLC,* 354) lives, then, if we trace Veblen's logic, is to camouflage the fact that male honor depends upon women. He observes in *Leisure Class* that "the good name of the household" is the woman's province, and the "honorific expenditure and conspicuous leisure by which this good name is chiefly sustained is therefore the woman's sphere" (180). Male status similarly depends on women in *Jennie Gerhardt* and is precarious for that very reason. When old Gerhardt confronts Senator Brander about courting Jennie, each wants to demonstrate that she will obey *his* will: the father to preserve his honor and the lover to maintain his worth. We catch again the glitter of the other man.

For Dreiser—especially in *Jennie Gerhardt*—and Veblen, women seem more essential than men, which means not only that women may be more easily trapped in essentializing scripts, but also that they form the most effective markers for male identity. I do not mean to suggest that Dreiser categorically denies men the benevolent qualities he assigns to Jennie or to argue that he saw all women as manifesting Jennie's disposition. *Twelve Men,* for instance, includes sketches of lives as altruistic as Jennie's—especially "A Doer of the Word" and "The Country Doctor." But Lester is depicted as ultimately more trapped by social convention, less able to live outside of it, than Jennie. In Veblen's social theory, let us recall, women have inherently a greater helping of *Homo sapiens's* redemptive traits of workmanship and the parental bent. (Thus Veblen reads the " 'New-Woman' movement" of the early twentieth century as an effort to "rehabilitate the woman's pre-glacial standing" [*TLC,* 356].) Similar to Lester Frank Ward's contention that "life begins as female" and "the male is therefore, as it were, a mere afterthought of nature," Veblen and Dreiser see men as incapable of standing erect without women to prop them up.[30]

To acknowledge that Dreiser and Veblen see women as more "essential" in this respect suggests that they anticipated modern insights into male gender formation. As anthropologist David D. Gilmore puts it, "Manhood ideologies force men to shape up on penalty of being robbed of their identity" and thus "manhood is a test in most societies."[31] A Veblenian analysis of *Jennie Gerhardt* not only reveals that the novel is one of Dreiser's most critical, but also how the ferociousness of approved models of masculinity may mask fundamental weakness. Although one side of Dreiser obviously likes to glorify hypermasculinity—as seen particularly in Frank Cowperwood—in *Jennie Gerhardt*, he critiques it.

A great deal of Lester's appeal as a character lies in his attempt to ignore the conventional tests of masculinity and construct manhood on his own terms, to withstand the practically compulsory expectations generated by his gender and class position. As Valerie Ross puts it, "Lester simultaneously signifies and problematizes conventional divisions between masculinity and femininity." "He really could not stand for this sort of thing any more," Lester thinks after returning from yet another dreary society party (*JG*, 146). Following the pattern identified by Leslie Fiedler for numerous male heroes in American literature, Lester rebels against both pecuniary and leisure-class values.[32] But in sharp contrast to Fiedler's paradigm, Dreiser imagines a *woman* providing the escape from the monotonous pressures of social convention: "*There* was someone who appealed to him," Lester thinks of Jennie (*JG*, 146). For many years he distances himself from the family business and withstands the pressures to marry, quietly defying his family and his entire class by his illicit relationship with Jennie. What is striking about how Dreiser depicts Lester's rebellion is that Jennie gives Lester the courage for it.

But, as Veblen remarks, "[t]he important and interesting institutions . . . [are those] of coercive control" (*POS*, 45). The classes at the

---

30. Lester Frank Ward, *Pure Sociology: A Treatise on the Origin and Spontaneous Development of Society*, 313, 314.
Other notably essentialized women in literary naturalism include Frank Norris's Hilma Tree of *The Octopus* and Moran of *Moran of the Lady Letty* (1898).
31. Gilmore, *Manhood in the Making*, 221, 220.
32. Valerie Ross, "Chill History and Rueful Sentiments in *Jennie Gerhardt*," 33; Fiedler, *Love and Death*.

top of the social hierarchy, having "a material interest in leaving things as they are," conserve destructive institutions to maintain their hegemony (*TLC*, 200). Since the upper classes retard change, innovation comes to be seen as "a lower-class phenomenon" and "vulgar." Consequently, "[i]nnovation is [seen as] bad form" (*TLC*, 200). In *Jennie Gerhardt*, Lester, "hedged about by the ideas of the conventional world," learns what little tolerance polite society has for change (*JG*, 243). Although his father's "attempt at coercion irritated Lester," he cannot resist it (*JG*, 278). Facing "the armed forces of convention, ... He could not fly in the face of it. He could not deliberately ignore its mandates" (*JG*, 368). The reason is the same that Veblen's analysis of institutional coercion predicts. Indeed, the language is virtually identical: "Society was made by the most conservative, who were almost invariably the most powerful. Their very conservatism was their power" (*JG*, 292). The narrator remarks that "[t]he conventions, in their way, appear to be as inexorable in their workings as the laws of gravitation and expansion" (*JG*, 283). Like Newland Archer in Edith Wharton's *The Age of Innocence*, who also worries about rebellion being "bad form," Lester cannot succeed as the romantic rebel he would like to be.[33] As a cultural critic, Lester fails even more dramatically than Bob Ames.

It is well known that for Dreiser, one of the most coercive institutions was marriage. The sense of inevitability that he generates during the final segment of *Jennie Gerhardt* illustrates his sense of the compulsory nature of the institution. Once the widowed Letty Pace Gerald reenters the novel as an appropriate wife for Lester, his defection from Jennie seems a foregone conclusion. As Veblen says, the "chief wife" in barbarian marriage usually has "gentle blood. . . . because a superior worth is felt to inhere in blood which has been associated with many goods and great power" (*TLC*, 54, 55). Letty, wealthy even before her first marriage to a millionaire, perfectly fits the bill. She entices Lester by behaving like a trophy-wife ("mak[ing] him feel as if he owned her" [*JG*, 319]) and promising to live vicariously for him ("enter[ing] on a dazzling social display for his sake" [*JG*, 377]). Her sickening pet name for Lester, "Mister-Master," pleases the barbarian in him (*JG*, 370).

After leaving Jennie, Lester experiences a "curious rejuvenation in [his] social and business spirit"; "authority" and "power" return to him (*JG*, 367). But there is really nothing curious about it. Having

---

33. Edith Wharton, *The Age of Innocence*, 8.

distanced himself from Jennie, the social pariah who is his moral superior, Lester's pecuniary instincts and male potency return to him. As he explains to Jennie after the fact, "I thought it good business to leave you" (*JG*, 392), and good for business it turns out to be. Lester grows corpulent, hires a chef at one hundred dollars a week, and takes on one of Veblen's favorite signs of conspicuous wealth, liveried servants. Lester's fall—and surely it is a fall—back into conventional life accentuates Dreiser's approbation of Jennie for remaining outside of the coercive conventions her lover spends much of his life fighting with little success.

Casting Out the Best

It is no wonder that Lester experiences Jennie's "care" as a "revelation" during their years together (*JG*, 221). Because of her commitment to service, her manifestation of what Veblen calls the instinct of workmanship and parental bent, Jennie embodies a better side of human nature than Dreiser usually shows. Dreiser intended that she be seen as the superior being in this novel, and while one is free to discount, disagree with, or deconstruct an author's purpose, that purpose remains worth establishing. "Truly she had played a finer part than he had," the narrator says of Jennie and Lester (*JG*, 369). As Dreiser describes Jennie's worth through Lester's consciousness, "She was charming . . . not strong or able in any of the ways the world measures ability, *but with something that was better*" (*JG*, 193, emphasis added).

Dreiser's cultural criticism emerges through the contrast between how "the world measures ability" and how he does. It surfaces also in how Dreiser uses other characters' responses to Jennie to indict the entire status system. At the center of the novel lies an obnoxious double standard, according to which the world judges Jennie. Not only is her common-law husband reclaimed socially and financially by the end of the novel, Lester himself censures his lover after discovering Vesta: "Senator Brander's child, he thought to himself. So that great representative of the interests of the common people was the undoer of her—a self-confessed washer-woman's daughter. A fine tragedy of low life all this was" (*JG*, 210). Lester's hypocrisy at this moment is astounding. Dreiser's social commentary matches Veblen's on this point: outcasts are likely to be society's most valuable members. He also exposes the perverse reasoning that condemns Jennie: because virtue places the needs of others before those of the self, people despise it. But as even Old Gerhardt finally comes to realize, "his outcast daughter was goodness itself" (*JG*, 344). Indeed, her goodness determines her status as social outcast.

The contrast between Jennie's and Lester's capacities for affection further discloses Dreiser's intent. "He liked her—loved her, perhaps, in a selfish way," the narrator remarks of Lester (*JG*, 187). In contrast, "all [Jennie's] attitude toward sex was bound up with love, tenderness, service." Thus, as Dreiser puts it in one of the descriptions most likely to grate on modern ears, Jennie is "a big woman, basically. . . . worthy of any man's desire" (*JG*, 72-73). The description is sexist, but it is not only that.[34] Jennie bestows her love without regard to the "wealth, social standing, personal force" of such interest to Dreiser's other romantically inclined characters. Because "her affections were not based . . . upon material considerations," she is "free from the taint of selfishness" (*JG*, 364). Again and again she demonstrates that "money was not the point of issue with her" (*JG*, 390). Her temperament quietly defies the pecuniary calculus and debased values around her.

Dreiser also admires Jennie because she manifests integrity, not invidiousness. She never, as the narrator remarks several times, loses her sense of proportion or her values (see *JG*, 167). As she prepares to make what the novel defines as the ultimate sacrifice, surrendering Lester to Letty Gerald, Jennie "felt humble," as if she were "holding handfuls of jewels that did not belong to her" (*JG*, 317). Lester himself realizes that Jennie is " 'peculiar. . . . She doesn't want much. She's retiring by nature and doesn't care for show' " (*JG*, 373).

Jennie's instinct of workmanship and parental bent set her apart from the values of the society in which she lives. The "archaic traits" that Veblen claimed could survive from the "peaceable cultural phase" define Jennie's personality: "conscience, including the sense of truthfulness and equity, and the instinct of workmanship, in its naïve, non-invidious expression" (*TLC*, 221). Lester, however, succumbs to the "predatory" traits of "self-seeking" and especially "clannishness" (*TLC*, 225). As Dreiser remarks, right after complaining of the "marvelously warped"

---

34. Dreiser also uses "big" as a positive descriptor for men. Archibald Kane (the father) is widely approved as "a big man" (*JG*, 137) while Lester, having a "bigger mental grasp of the subtleties which compose life" than his brother, is considered "the biggest one in that family" (*JG*, 137, 120).

and "radically wrong" values that are socially dominant, "As yet, we are dwelling a most brutal order of society" (*JG,* 92, 93). In *Jennie Gerhardt,* Dreiser does what Veblen does consistently throughout his corpus: sympathize with the despised, valorize the marginal, and trace the reasons why a pecuniary civilization belittles the virtuous poor.

Sentimentalism and Cultural Criticism

Such indications of Jennie's innate moral superiority take us into the heart of the logic of sentimental fiction. *Jennie Gerhardt's* historically poor ranking in the Dreiser canon forms a chapter in the long-standing critical repugnance against sentimentalism. Much of the critical discussion of *Jennie Gerhardt* argues either that Dreiser succumbs to sentimentalism or finally "escapes" it, thereby redeeming his novel. In particular, objections to Jennie's character generally concentrate on her sentimental outlines. Mordecai Marcus, for instance, sees the "sometimes superficial and sentimentalized portrayal of Jennie" as the novel's "greatest weakness," while Lawrence Hussman notes that Jennie's "sacrifices for her family and her lovers are so conspicuous that they constitute a nearly fatal flaw in the novel." Hussman concludes that Jennie is "one dimensional," of "doughlike consistency." The reception of *Jennie Gerhardt* points to the wider issue that Amy Kaplan refers to as the "problem of Dreiser's sentimentalism." There is indeed a problem, but I would locate it more within the critical discourse than in Dreiser's novels.[35]

The problem is twofold. First whereas the question of Dreiser's sentimentalism in *Sister Carrie* has received sophisticated treatment, most commentators who broach *Jennie Gerhardt* take it for granted that sentimentalism is second-rate. Leslie Fiedler exemplifies this attitude when he argues that the "fundamental flaw" in all of Dreiser's novels

---

35. Mordecai Marcus, "Loneliness, Death, and Fulfillment in *Jennie Gerhardt,*" 61; Hussman, *Dreiser and His Fiction,* 64, 65 (Hussman renews his objections in "Jennie One-Note: Dreiser's Error in Character Development"); Kaplan, *Social Construction,* 142. Kaplan argues for "the interpenetration of realism and sentimentality" in *Sister Carrie* (ibid., 140), but characterizes sentimentalism as escapist (ibid., 144, 159). She ties the desire for "revolt" to sentimentalism in *Sister Carrie,* but sees a fundamental "gap between desire and social power that reopens the space for sentimentality" (148, 151). I am arguing that the sentimentalism in *Jennie Gerhardt* is instrumental in Dreiser's cultural criticism.

is their being "sentimental rather than tragic."[36] It is easy to see what is wrong with this line of reasoning. As Nina Baym explains, the "term 'sentimental' is often a term of judgment rather than description and the judgment it conveys is of course adverse."[37]

The second problem with critical discussion of sentimentalism in *Jennie Gerhardt* follows from the way supposedly descriptive literary-historical labels take on pejorative connotations. Most commentators assume an innate opposition exists between realism/naturalism and sentimentalism when the two literary modes have much in common.[38] It is difficult to imagine a thorough analysis of *Jennie Gerhardt* that does not make use of such tropes central to sentimental fiction as the moral superiority of woman, the centrality of the home (always a primary issue with Dreiser), seduction, emotionalism, "imperiled womanhood," "salvation through motherly love," an "ethic of sacrifice" or "ethic of submission."[39] *Jennie Gerhardt*'s sharing so many characteristics with

---

36. Fiedler, *Love and Death*, 250. Similar assessments include Pizer, *Novels*, 107; Marcus, "Loneliness," 73; Daryl C. Dance, "Sentimentalism in Dreiser's Heroines Carrie and Jennie," especially 142; Gerber, *Theodore Dreiser*, 43. More judicious discussion of sentimentalism in *Sister Carrie* can be found in Kaplan, *Social Construction*; Michaels, *Gold Standard*; and Petrey, "Language of Realism."

It is commonly held that sentimentalism is at odds with realism and/or naturalism, for example, in Carol A. Schwartz, "*Jennie Gerhardt:* Fairy Tale as Social Criticism," 17; and Susan Wolstenholme, "Brother Theodore, Hell on Women," 248. The definitive statement of the alleged dichotomy in Dreiser is Petrey's treatment of the "two irreconcilable styles" in "The Language of Realism," 102. Several of the essays in West's collection, *Dreiser's "Jennie Gerhardt"*—particularly Ross's "Chill History"—reassess sentimentalism in *Jennie Gerhardt*.

37. Nina Baym, *Woman's Fiction: A Guide to Novels by and about Women in America, 1820-70*, 24.

38. Nina Baym, "The Rise of the Woman Author," 292; Habegger, *Gender, Fantasy, and Realism in American Literature;* Kaplan, *Social Construction;* Howard, *Form and History,* 175.

The characteristic plots of sentimentalism and naturalism may also share a fundamental similarity. In *Sensational Designs: The Cultural Work of American Fiction 1790-1860,* Jane Tompkins writes that the typical plot of sentimental fiction "does not unfold according to Aristotelian standards of probability, but in keeping with the logic of a preordained design" (135). The plots of naturalist novels frequently strike readers as inevitable as well.

In "*Jennie Gerhardt:* Naturalism Reconsidered," Judith Kucharski asks that we "pretend for a moment that we do not 'know,' as we think we do, that Dreiser is a 'naturalist' or a 'pessimistic realist' or a 'mechanistic determinist' " (17). We can then see that in *Jennie Gerhardt,* Dreiser "critiqu[es] the naturalistic philosophy he supposedly endorsed" (ibid., 24).

39. "Imperiled womanhood" comes from Lora Romero, "Domesticity and Fiction," 112; the other three quotations from Tompkins, *Sensational Designs,* 125, 128, 161. My characterization of sentimental tropes—which draws particularly from Romero,

sentimentalism is evidence not that the novel betrays realist/naturalist codes but that we need to rethink the boundaries critics have constructed around literary modes and also more scrupulously to monitor how aesthetic preference often masquerades as literary history. Let us not pretend to be without preferences, but let us own up to them.

Jennie Gerhardt's sentimentalism is not something that Dreiser needs to escape. Nina Baym remarks, "If critics ever permit the woman's novel to join the main body of 'American' literature, then all our theories about American fiction, from Richard Chase's 'romance' to Richard Poirier's 'world elsewhere' to Carolyn Heilbrun's 'masculine wilderness' will have to be radically revised." The explosion of interest in sentimental fiction that Baym pioneered has indeed precipitated reconsideration of established theories about American fiction. If Dreiser critics look more carefully at *Jennie Gerhardt*—particularly its sentimentalism and gender essentialism—a major reassessment of his work may result. As Joseph Epstein provocatively remarks, "the character of Jennie Gerhardt stands as a refutation of the theory that is supposed to lie behind much of Theodore Dreiser's fiction."[40] To treat *Jennie Gerhardt*, sentimentalism and all, without condescension requires reconceptualizing the entire Dreiser canon. One way to begin doing so is to take seriously the cultural criticism Dreiser advances in his second novel.

The gains to be had from reconsidering *Jennie Gerhardt* in conjunction with a recuperative analysis of sentimentalism become particularly apparent in light of Jane Tompkins's influential model. Tompkins's magisterial refutation of the charge of escapism is that sentimental literature, a "political enterprise, halfway between sermon and social theory," has "designs on the world." Sentimental fiction, she argues, far from avoiding political reality, discloses power relations and proposes an alternative theory of power, "operat[ing] . . . according to a principle of reversal whereby what is 'least' in the world's eyes becomes

---

Tompkins, Baym's *Woman's Fiction* and "The Rise of the Woman Author," Ann Douglas's *The Feminization of American Culture*, and Phillip Brian Harper's "Fiction and Reform II"—is not intended to suggest uniform agreement among critics. Indeed, in *Woman's Fiction*, Baym disputes the relevance of the seduction trope that others see as central to sentimentalism (26), while in "Bio-Political Resistance in Domestic Ideology and *Uncle Tom's Cabin*," Romero criticizes the oppositional model on which Tompkins's entire analysis rests.

40. Baym, *Woman's Fiction*, 36–37; Joseph Epstein, "A Great Good Girl: Dreiser's *Jennie Gerhardt*," 19.

'greatest.' "[41] The reversal of dominant social values and social power that Tompkins identifies as the axiological core of sentimental fiction informs the logic of *Jennie Gerhardt*, as well as Veblen's lionization of the instinct of workmanship and the parental bent.

Tompkins's analysis of the oppositional politics of sentimentalism makes it easy to refute Leslie Fiedler's wrongheaded charge that Dreiser's "famous determinism is essentially sentimental at root, amounting effectively to little more than the sob of exculpation: 'Nobody's fault!' "—truly an astounding misreading.[42] Because he begins with the premise that sentimentalism means mushy escapism, Fiedler misconstrues Dreiser's complex and scathing presentation of social problems as an inability to pass judgment on them. Fiedler occludes the considerable distance between depicting social problems—appropriate work for a cultural critic to do, in fiction or in social theory—and resolving them.

Dreiser's treatment of "My Brother Paul" in *Twelve Men* provides an extended look at his views of sentimentalism. Dreiser's backhanded compliment of his brother Paul's hugely successful songs exemplifies the ambivalence that permeates the sketch. Paul Dresser (who had Americanized the spelling of the family name) wrote tunes that Dreiser called "pale little things . . . mere bits and scraps of sentiment and melodrama . . . most asinine sighings over home and mother and lost sweethearts and dead heroes such as never were in real life, and yet with something about them . . . which always appealed to me intensely and must have appealed to others."[43] Right after voicing some of the most common complaints against sentimentalism, Dreiser admits its power nonetheless.

But even more than chronicling Paul's career as a sentimental balladeer, "My Brother Paul" registers a personality—and herein lies the real power. Like Jennie Gerhardt, "[s]ympathy was really [Paul's] outstanding characteristic." "[G]enerous to the point of self-destruction," Paul was distinguished by "his really great heart." As he does with Jennie,

---

41. Tompkins, *Sensational Designs*, 126, 125, 162-63. See also Banta in *Taylored Lives*: "Sentimentality immediately calls up images of the conservative evasion of direct confrontations with social injustice, but it can just as surely color the practices of would-be radicals" (26). In *Tales*, Hapke affirms that *Jennie Gerhardt* is both sentimental and radical, but sees the two tendencies as competing (83).

42. Fiedler, *Love and Death*, 249.

43. Theodore Dreiser, "My Brother Paul," 97. See also Kaplan, *Social Construction*, 117-22, 137-39.

Dreiser compares Paul numerous times to mothers in general and to their mother in particular. The biographical sketch also records a prototype for the emotional scene in *Jennie Gerhardt* when Senator Brander sends Christmas gifts to the entire Gerhardt family; in "My Brother Paul" the recipient is sister "E—" (Emma). Dreiser's own response to this event is telling: "There were tears in my eyes, too. One couldn't resist him."[44]

As Dreiser's response confirms what Jane Tompkins calls sentimental power, it indicates also his refashioning of that power to include men. For although the fictional character in whom he invests the most sentimental power is his essentialized woman, Jennie Gerhardt, as a male author, Dreiser needs to ascertain that sentimental affect and masculinity are not inherently opposed. H. L. Mencken made the intriguing comment that Dreiser is "perhaps the only American novelist who shows any sign of being able to feel profoundly" and because "his ideas always seem to be deduced from his feelings," Mencken concluded that Dreiser's "talent is essentially feminine." However dubious Mencken's typecasting by gender, he rightly draws attention to the emotive power of Dreiser's novels. More recently, Michael Davitt Bell has argued that Dreiser "was not stereotypically 'masculine'" although "a prominent function of claiming to be a realist or a naturalist in this period was to provide assurance to one's society and oneself that one was a 'real' man rather than an effeminate 'artist.'"[45] Dreiser's indulgence in sentimentalism, facilitated by the example of Paul, helped him to move beyond masculine stereotypes. Paul was successful in the eyes of the world and highly attractive to women (and, Dreiser remarks, Paul was the only man he knew more highly sexed than himself [*D*, 154]). With these

---

44. Dreiser, "My Brother Paul," 82, 79, 77, 88, 89. Dreiser's comparisons of Paul and their mother can be seen in ibid., 78, 85, 94, 103.

45. Mencken, "Adventures," 796; H. L. Mencken, "Theodore Dreiser," 787; Bell, *The Problem of American Realism*, 152, 6. Bell explains Dreiser's position by reading hypermasculinity as "in part a symptom of upper- or upper-middle-class anxiety that privilege led to a sapping of virility, and the problems of privilege hardly played a role in Dreiser's early years" (ibid., 153). Dreiser's class status, in other words, exempts him from the cult of masculinity.

According to Vera Dreiser in *My Uncle Theodore*, 30–37, the novelist was "permanently damag[ed]" by his mother, Sarah. She charges Sarah with doing "great damage to [Theodore's] masculinity," causing him to develop an "almost feminine sensitivity," which was, in turn, responsible for his creativity (33, 37). Vera Dreiser's partiality is clear, but her family-based study offers a corrective for the tendency of Dreiser—and many of his critics—to venerate Sarah Dreiser.

credentials both worldly and erotic, Paul constitutes for the younger brother a model of a man who was successful—and had sentimental power.

Dreiser occasionally defined himself as a "sentimentalist," even an "excitable and high-flown sentimentalist" (*HH,* 77; *AL,* 102), but his relationship with sentimentalism was conflicted. "My Brother Paul" suggests what was at stake for Dreiser. In one of the most emotionally charged sections, he records collaborating with his brother on the popular song, "On the Banks of the Wabash." As Dreiser portrays the event, he offhandedly suggests that Paul write a song about a river, and Paul then prods him into writing the words himself. Dreiser capitulates, but he resists: "I was convinced that this work was not for me and that I was rather loftily and cynically attempting what my good brother would do in all faith and feeling." While Dreiser plays cynical realist to Paul's emotional sentimentalist, he invokes what is perhaps the greatest complaint that has been lodged against sentimentalism: the possibility that its strong emotionalism is contrived and inauthentic. So James Baldwin indicts sentimentalism (his particular target is *Uncle Tom's Cabin*) for its "mark of dishonesty, the inability to feel." Yet Dreiser, despite his misgivings about the possible inauthenticity of sentimentalized emotion, writes on. Dreiser discovers that the questionable emotions become so real as to threaten to engulf him: "I still protested weakly, but in vain. [Paul's] affection was so overwhelming and tender that it made me weak."[46] Suddenly Dreiser seems to be playing fair maiden to Paul's tender seducer. Theodore Dreiser—seduced by sentiment.

This episode indicates how thoroughly entangled Dreiser could become with sentimental affect. What Tompkins refers to as the "emotional exhibitionism" of sentimentalism is something that many people fear—including, at times, Dreiser himself, and I think many of his critics. As Saul Bellow remarked long ago, "the criticism of Dreiser as a stylist at times betrays a resistance to the feelings he causes readers to suffer.

---

46. Dreiser, "My Brother Paul," 101; Baldwin, "Everybody's Protest Novel," 28; Dreiser, "My Brother Paul," 104.

One can, of course, align Dreiser with emotion and still find the combination objectionable. Zelda Fitzgerald, for one: "I have an intense distaste for the melancholy aroused in the masculine mind by such characters as Jenny [sic] Gerhardt, [Willa Cather's] Antonia and Tess [of the D'Urbervilles]. Their tragedies, redolent of the soil, leave me unmoved" (Fitzgerald quoted in Elizabeth Kaspar Aldrich, " 'The most poetical topic in the world,' " 138).

If they say he can't write, they need not express those feelings."[47] But inducing strong affective responses in readers helps Dreiser to produce cultural criticism.

In *Jennie Gerhardt,* Dreiser uses Lester to indicate the difference between vulgar and legitimate sentimentalism. Lester, who wants never to appear "mawkishly sentimental" (*JG,* 128), is appalled by the Chicago newspapers' account of his ménage with Jennie. One paper does seem to combine elements that might be labeled sentimental and realistic— "running over with sugary phrases, but still, and in spite of itself, with the dark, sad facts looming up in the background" (*JG,* 286)—but the emotionalism is insincere, and the resulting account distorts reality. Nor are headlines such as "Sacrifices Millions for His Servant-Girl Love," which presumably boil the story down to its key facts, particularly accurate. Lester sees it as an "asinine attempt to sugar over the true story, and it made him angry" (*JG,* 285, 290). Such sugarcoating is the antithesis of what Dreiser achieves by fusing realism/naturalism and sentimentalism to produce cultural criticism in *Jennie Gerhardt.*

Unlike Dreiser, Veblen restrains emotional effects in his writings; when he wants to jar readers, he goes for the jugular rather than the heart. Veblen's works demonstrate none of the emotional content of sentimentalism. But there exists an important, and neglected, softer side to Veblen that engages in some curious relays with the sentimental. Joseph Dorfman's biography of Veblen focuses more on historical context than on psychological portraiture, but Veblen's personality begins to emerge in an important document brought to light by Rick Tilman, which he calls "[p]erhaps the most focused analysis of Veblen the man." Jacob Warshaw, professor of romance languages at the University of Missouri–Columbia, and Veblen's friend during his seven years there, criticizes Dorfman's description of Veblen's personality. According to Warshaw, Veblen was not, "in his heart, the suave, imperturbable, sphinx-like character who stands out in Dorfman. He struck me rather as a man of spontaneous passions." Furthermore, writes Warshaw, "under the surface, Veblen was highly emotional. . . . To realize this emotional quality in Veblen is, it seems to me, to get another light on his ideas and projects, to forgive him most of his cynicism, and to

---

47. Tompkins, *Sensational Designs,* 132; Saul Bellow quoted in Epstein, "Great Good Girl," 15.

find him a more sympathetic and human individual than he is usually credited with being."[48]

Veblen's works engage in a latent as well as a blatant dialogue with the ethic of sentimentalism. The latent component can be seen in Veblen's construction of sentimental savages: virtue unrewarded, a female-centered social order, women whisked away by predatory seducers. Most significantly aligning him with the sentimental ethic, Veblen's positive values emerge clearly in this domain. Some of the most sentimental passages ever written by Veblen show the interweaving of the instinct of workmanship and the parental bent. For instance, he hypothesizes that during the savage era,

> the scheme of life of the crops and flocks is . . . a scheme of fecundity, fertility and growth. But these matters, visibly and by conscious sentiment, pertain in a peculiarly intimate sense to the women. They are matters in which the sympathetic insight and fellow-feeling of womankind should in the nature of things come very felicitously to further the propitious course of things. . . . There is a magical congruity of great force as between womankind and the propagation of growing things. (*IOW*, 92-93)

This "magical congruity" that Veblen posits between Neolithic women and nature derives from the startling premise that they communicate wordlessly with plants and animals: "It is all the more evident that communion with these wordless others should fall to the women, since the like wordless communion with their own young is perhaps the most notable and engaging trait of their own motherhood" (*IOW*, 94). Surely Veblen seems more sentimental than splenetic here.

Veblen's explicit comments on sentimentalism are, typically enough, circuitous. Although three novels by the American sentimentalist E. D. E. N. Southworth turned up in the Veblen family library, he does not, of course, say anything about such literature in his writings.[49] Veblen does, however, concern himself with the cultural construction of territory that is demarcated as sentimental. Most commentators on sentimentalism, its detractors and defenders alike, concur that it

---

48. Tilman, *Thorstein Veblen and His Critics*, 6; Jacob Warshaw, "Recollections of Thorstein Veblen," 3.

49. The Southworth novels are *The Changed Brides, Fair Play,* and *The Fatal Marriage,* all of them now at the Thorstein B. Veblen Collection at the Carleton College Archives.

celebrates behaviors traditionally coded as feminine. Veblen's distinctive contribution to the discussion of gender and sentimentalism is to demonstrate how approved *manly* behaviors are themselves "sentimental," thus furthering his exposé of dominant cultural values.

Veblen's relocation of sentimentalism within the male sphere constitutes a powerful, if neglected, component of his cultural criticism. Occasionally he uses the word *sentimental* in (or at least close to) its normative meaning, as with the claim that "Business management has a chance to proceed on a temperate and sagacious calculation of profit and loss, untroubled by sentimental considerations of human kindness or irritation or of honesty" (*TBE,* 53). Here Veblen quickly extends from the familiar association of sentimentalism with human kindness, in contrast to the unemotional calculations of business, to an ironically charged characterization of business as dishonest.

More frequently, however, Veblen locates sentimentalism within the manly realm of business rather than contrasting them. He argues, for instance, that the belief in businessmen "coördinating . . . industrial processes with a view to economics of production and heightened serviceability," an idea that Veblen's entire corpus strives to disprove, "has a great sentimental value and is useful in many ways" (*TBE,* 41)— useful, that is, for maintaining business as usual. Veblen reiterates the point in *The Engineers and the Price System,* arguing that there is no foreseeable revolution in the United States because "[b]y settled habit, the American population is quite unable to see their way to entrust any appreciable responsibility to any other than business men. . . . This *sentimental deference* of the American people to the sagacity of its business men is massive, profound, and alert" (139, emphasis added). Veblen emphasizes that it is, specifically, American males who demonstrate this type of sentimental deference. Indeed, he sharply remarks, there is in the United States "a sentimental conviction that pecuniary success is the final test of manhood" (*HLA,* 82).

Commentators may have overlooked this dimension of Veblen's cultural criticism because we are not accustomed to look for sentimentalism in discussions of the traditionally masculine territories that he analyzes, such as business, sports, and nationalism. But besides exposing the "sentimental penchant for large figures" characteristic of business (*HLA,* 115), Veblen extends his astute analysis of the sentimentalism of American males into other approved masculine territories. Sports, particularly intercollegiate games, again indicate "sentimental rivalry" between teams (*HLA*, 235). And in the same way that deference to business interests helps to maintain the status quo, so do other signs of unthinking loyalty, such as patriotism. Indeed, "[p]atriotism or loyalty is a frame of mind and rests on a sentimental adhesion to certain idealistic aspirations. . . . [which] will commonly not bear analysis" (*IG*, 212). Veblen thus agrees with the critics of sentimentalism that the mode is intellectually flaccid—but he assigns the illogic to males. He consistently denounces clannish behaviors, the tendency of people (particularly, he argues, male people) to band together. Such instances of "group solidarity," mockeries of the real communalism associated with the parental bent and instinct of workmanship, betray "sentimental sophistry" (*IG*, 47).[50]

The instinct of workmanship and parental bent make up a positive side of Veblen's thinking that scholars have frequently noted, but almost as frequently downplayed. He strategically assigns women a larger component of these positive traits, and he uses the contrast to critique the excesses of the capitalist order.

As the instinct of workmanship and parental bent form an important counterpoint to Veblen's often bleak view of human nature, so is Jennie Gerhardt—both character and novel—an exceptional Dreiser creation. Far from seeming a "fallen woman," Jennie rises above the society that condemns her. Yet Jennie's initial "sin" in the eyes of the barbaric world was, after all, having sex outside of marriage, a fact linking her with Carrie Meeber, Clyde Griffiths, Roberta Alden, and even more significantly, with Dreiser's personifications of intransigence, Eugene Witla and Frank Cowperwood. When it comes to the cultural work Dreiser does with his characters—the work of cultural criticism—there is surprisingly little difference between Jennie Gerhardt and Frank Cowperwood, who

---

50. Veblen frequently uses the word *sentiment*, its seemingly neutral meaning shading into what he means by "sentimental." For instance, when Veblen mentions the "popular sentiment in favour of sports" or the "recrudescence of anthropomorphic sentiment" (*TLC*, 270, 373), he seems to be critiquing the approved masculine activities. A particularly interesting use of the word occurs in one of Veblen's assaults on the reigning orthodoxies in economics: "By a bold metaphor—a metaphor sufficiently bold to take it out of the region of legitimate figures of speech—the gains that come to enterprising business concerns by such monopolistic enhancement of the 'total effective utility' of their products are spoken of as 'robbery,' 'extortion,' 'plunder'; but the theoretical complexion of the case should not be overlooked by the hedonistic theorist in the heat of outraged sentiment" (*POS* 216). Hedonistic "sentiment," it seems, obscures rational thought.

in other respects seem antithetically opposed. Dreiser depicts Jennie and Cowperwood as consummate outsiders, both maintaining their integrity without bowing to ceremonial considerations or institutional pressures. Manifesting the traits Veblen calls the instinct of workmanship and parental bent, Jennie's kindness, industry, and solicitude for others help Dreiser to advance what Mencken called "a criticism and an interpretation of life."[51]

Jennie Gerhardt confirms Mencken's point that Dreiser can interpret American life in fictional form while rendering the same material into cultural criticism. As the ideas of Veblen—at times parallel and at times complementary—help to draw out Dreiser's interventions in several phases of cultural interpretation, so do Dreiser's writings help to spotlight Veblen's reflections on a wide range of questions that continue to preoccupy analysts of the American scene. Their musings on the role of consumption in modern life; on the place of business in the American imagination; on the complex relays between gender, class, and power; and on the role of the critical intellectual in presenting these issues to the public, all demonstrate the fine art of cultural criticism. Neither Veblen nor Dreiser can be contained in any disciplinary or descriptive boxes we can contrive; both of them deserve recognition as remarkably multifaceted—and prescient—saboteurs of the status quo.

---

51. Mencken, "A Novel of the First Rank," 741.

BIBLIOGRAPHY

Manuscript Collections

Rasmus Bjorn Anderson Papers, 1823-1936. State Historical Society of Wisconsin, Madison.
Joseph Dorfman Collection. Butler Library, Columbia University.
Theodore Dreiser Collection. Van Pelt-Dietrich Library, University of Pennsylvania.
Jacques Loeb Papers. Manuscript Division, Library of Congress.
Andrew Veblen Papers. Minnesota Historical Society, St. Paul.
Thorstein B. Veblen Collection. Carleton College, Northfield, Minnesota.
Jacob Warshaw Papers, 1910-1944. Western Historical Manuscript Collection, Columbia, Missouri.

Books and Articles

Aaron, Daniel. *Men of Good Hope: A Story of American Progressives.* New York: Oxford University Press, 1951.
Adorno, Theodor. "Veblen's Attack on Culture." In *Prisms,* trans. Samuel and Shierry Weber, 75-94. Cambridge: MIT Press, 1981.
Agnew, Jean-Christophe. "The Consuming Vision of Henry James." In *The Culture of Consumption,* ed. Richard Wightman Fox and T. J. Jackson Lears, 65-100. New York: Pantheon, 1983.
———. "A House of Fiction: Domestic Interiors and the Commodity Aesthetic." In *Consuming Visions: Accumulation and Display*

of Goods in America, 1880-1920, ed. Simon J. Broner, 133-55. New York: Norton, 1989.

Albertine, Susan. "Triangulating Desire in *Jennie Gerhardt.*" In *Dreiser's "Jennie Gerhardt": New Essays on the Restored Text,* ed. James L. W. West III, 63-74. Philadelphia: University of Pennsylvania Press, 1995.

Aldrich, Elizabeth Kaspar. " 'The Most Poetical Topic in the World': Women in the Novels of F. Scott Fitzgerald." In *Scott Fitzgerald: The Promises of Life,* ed. A. Robert Lee, 131-56. New York: St Martin's, 1989.

Anesko, Michael. *"Friction with the Market": Henry James and the Profession of Authorship.* New York: Oxford University Press, 1986.

Arendt, Hannah. *The Human Condition.* Chicago: University of Chicago Press, 1958.

Ayres, C. E. "Veblen's Theory of Instincts Reconsidered." In *Thorstein Veblen: A Critical Reappraisal,* ed. Douglas F. Dowd, 25-38. Ithaca: Cornell University Press, 1958.

Bakhtin, M. M. *The Dialogic Imagination.* Ed. Michael Holquist. Trans. Caryl Emerson and Michael Holquist. Austin: University of Texas Press, 1981.

Baldwin, James. "Everybody's Protest Novel." In *The Price of the Ticket: Collected Nonfiction, 1948-1985,* 27-33. New York: St Martin's Press, 1985.

Banta, Martha. *Taylored Lives: Narrative Productions in the Age of Taylor, Veblen, and Ford.* Chicago: University of Chicago Press, 1993.

Barrineau, Nancy Warner. " 'Housework Is Never Done': Domestic Labor in *Jennie Gerhardt.*" In *Dreiser's "Jennie Gerhardt": New Essays on the Restored Text,* ed. James L. W. West, 127-35. Philadelphia: University of Pennsylvania Press, 1995.

Bartley, Russell H., and Sylvia E. Bartley. "In Search of Thorstein Veblen: Further Inquiries into His Life and Work." *International Journal of Politics, Culture and Society* 11:1 (fall 1997): 129-74.

Bartley, Russell H., and Sylvia E. Yoneda. "Thorstein Veblen on Washington Island: Traces of a Life." *International Journal of Politics, Culture and Society* 7:4 (summer 1994): 589-614.

Baudrillard, Jean. "The Precession of Simulacra." In *Art after Modernism: Rethinking Representation,* ed. Brian Wallis, 253-81. New York: New Museum of Contemporary Art, 1984.

———. *Simulations*. Trans. Paul Fos, Paul Patton, and Philip Beitchman. New York: Semiotext(e), 1983.

Baym, Nina. "The Rise of the Woman Author." In *Columbia Literary History of the United States*, gen. ed. Emory Elliott, 289-305. New York: Columbia University Press, 1988.

———. *Woman's Fiction: A Guide to Novels by and about Women in America, 1820-70*. 2d ed. Urbana and Chicago: University of Illinois Press, 1993.

Beer, Gillian. *Darwin's Plots: Evolutionary Narrative in Darwin, George Eliot, and Nineteenth-Century Fiction*. London: Routledge and Kegan Paul, 1983.

Bell, Michael Davitt. *The Problem of American Realism: Studies in the Cultural History of a Literary Idea*. Chicago: University of Chicago Press, 1993.

Bensman, Joseph. "The Aesthetics and Politics of Footnoting." *International Journal of Politics, Culture, and Society* 1:3 (spring 1988): 443-70.

Benstock, Shari, and Suzanne Ferriss, eds. *On Fashion*. New York: Rutgers University Press, 1994.

Berger, John. *Ways of Seeing*. New York: Penguin, 1977.

Bersani, Leo. *A Future for Astyanax: Character and Desire in Literature*. Boston: Little, Brown & Co Inc, 1976.

Berthoff, Warner. "Culture and Consciousness." In *Columbia Literary History of the United States*, gen. ed. Emory Elliott, 482-98. New York: Columbia University Press.

Bledstein, Burton. *The Culture of Professionalism: The Middle Class and the Development of Higher Education in America*. New York: Norton, 1976.

Borus, Daniel H. *Writing Realism: Howells, James, and Norris in the Mass Market*. Chapel Hill: University of North Carolina Press, 1989.

Bourne, Randolph. "The Art of Theodore Dreiser." Reprinted in *Critical Essays on Theodore Dreiser*, ed. Donald Pizer, 13-14. Boston: G. K. Hall & Co., 1981.

Bowlby, Rachel. *Just Looking: Consumer Culture in Dreiser, Gissing and Zola*. New York: Methuen, 1985.

———. "Soft Sell: Marketing Rhetoric in Feminist Criticism." In *The Sex of Things: Gender and Consumption in Historical Perspective*, ed. Victoria De Grazia and Ellen Furlough, 381-88. Berkeley: University of California Press, 1996.

Brandon, Craig. *Murder in the Adirondacks: "An American Tragedy" Revisited.* Utica, N.Y.: North Country Books, 1981.

Bratton, Daniel Lance. "Conspicuous Consumption and Conspicuous Leisure in the Novels of Edith Wharton." Ph.D. diss., University of Toronto, 1983.

Brennan, Stephen. "*The Financier:* Dreiser's Marriage of Heaven and Hell." *Studies in American Fiction* 19 (1991): 55-69.

———. "The Publication of *Sister Carrie:* Old and New Fictions." *American Literary Realism* 18:1 and 2 (spring and autumn 1985): 55-68.

Brown, Doug. "An Institutionalist Look at Postmodernism." *Journal of Economic Issues* 25:4 (December 1991): 1089-104.

Bush, Paul D. "The Theory of Institutional Change." Reprinted in *Economics of Institutions,* ed. Geoffrey M. Hodgson, 510-51. Brookfield, Vt.: Edward Elgar, 1993.

Byrne, David. "Angels." *David Byrne.* Sire Records, Warner Brothers, 1994.

Campbell, Colin. "Conspicuous Confusion? A Critique of Veblen's Theory of Conspicuous Consumption." *Sociological Theory* 13:1 (March 1995): 37-47.

———. "Romanticism and the Consumer Ethic: Intimations of a Weber-Style Thesis." *Sociological Analysis* 44:4 (winter 1983): 279-95.

Cappetti, Carla. *Writing Chicago: Modernism, Ethnography, and the Novel.* New York: Columbia University Press, 1993.

Caraher, Catherine Ann. "Thorstein Veblen and the American Novel." Ph.D. diss., University of Michigan, 1966.

Casciato, Arthur D. "How German is *Jennie Gerhardt?*" In *Dreiser's "Jennie Gerhardt": New Essays on the Restored Text,* ed. James L. W. West, 167-82. Philadelphia: University of Pennsylvania Press, 1995.

Cassuto, Leonard. "Dreiser's Ideal of Balance." In *Dreiser's "Jennie Gerhardt": New Essays on the Restored Text,* ed. James L. W. West III, 51-62. Philadelphia: University of Pennsylvania Press, 1995.

Chase, Richard. *The American Novel and Its Tradition.* Garden City, N.Y.: Doubleday Anchor, 1957.

Clark, John Maurice. "Thorstein Bunde Veblen, 1857-1929." 1929. Reprinted in *Essays, Reviews, and Reports,* by Thorstein Veblen.

Ed. Joseph Dorfman, 595-600. Clifton, N.J.: Augustus M. Kelley, 1973.

Conder, John J. *Naturalism in American Fiction: The Classic Phase.* Lexington: University Press of Kentucky, 1984.

Conroy, Stephen. "Thorstein Veblen's Prose." *American Quarterly* 20:3 (fall 1968): 605-15.

Corkin, Stanley. "*Sister Carrie* and Industrial Life: Objects and the New American Self." *Modern Fiction Studies* 33:4 (winter 1987): 605-19.

Cowley, Malcolm. "A Natural History of American Naturalism." Reprinted in *Documents in Modern Literary Realism,* ed. George J. Becker, 429-51. Princeton: Princeton University Press, 1963.

———. "Naturalism in American Literature." In *Evolutionary Thought in America,* ed. Stow Persons, 300-333. New Haven: Yale University Press, 1950.

Cowley, Malcolm, and Bernard Smith, eds. *Books That Changed Our Minds.* New York: Doubleday, Doran & Company, 1940.

"Current Fiction." Reprinted in *Theodore Dreiser: The Critical Reception,* ed. Jack Salzman, 121-22. New York: David Lewis, 1972.

Dance, Daryl C. "Sentimentalism in Dreiser's Heroines Carrie and Jennie." *CLA Journal* 14 (December 1970): 127-42.

Darwin, Charles. *The Descent of Man,* and *Selection in Relation to Sex.* 1871. Reprint, Princeton: Princeton University Press, 1981.

———. *The Origin of Species by Means of Natural Selection, or The Preservation of Favoured Races in the Struggle for Life.* 1859. Reprint, New York: Penguin, 1985.

Debord, Guy. *The Society of the Spectacle.* Detroit: Black & Red, 1977.

Degler, Carl. *In Search of Human Nature: The Decline and Revival of Darwinism in American Social Thought.* New York: Oxford University Press, 1991.

De Grazia, Victoria, and Ellen Furlough, eds. *The Sex of Things: Gender and Consumption in Historical Perspective.* Berkeley: University of California Press, 1996.

de Lauretis, Teresa. *Alice Doesn't: Feminism, Semiotics, Cinema.* Bloomington: Indiana University Press, 1984.

Deleuze, Gilles. *The Logic of Sense.* Trans. Mark Lester. New York: Columbia University Press, 1990.

Dente, Leonard. *Veblen's Theory of Social Change.* New York: Arno Press, 1977.

Dewey, John. "The Influence of Darwinism on Philosophy." 1909. Reprinted in *The Influence of Darwin on Philosophy and Other Essays in Contemporary Thought*, 1-19. New York: Peter Smith, 1951.
Dickstein, Morris. *Double Agent: The Critic and Society.* New York: Oxford University Press, 1992.
Diggins, John P. *The Bard of Savagery: Thorstein Veblen and Modern Social Theory.* New York: Seabury Press, 1978.
———. "Dos Passos and Veblen's Villains." *The Antioch Review* 23 (1963): 485-500.
———. *The Promise of Pragmatism: Modernism and the Crisis of Knowledge and Authority.* Chicago: University of Chicago Press, 1994.
———. "A Radical with Authority." *The Chronicle of Higher Education* 13 (November 1, 1976): 32.
Doane, Mary Ann. *The Desire to Desire: The Woman's Film of the 1940s.* Bloomington: Indiana University Press, 1987.
Dorfman, Joseph. "Background of Veblen's Thought." In *Thorstein Veblen: The Carleton College Veblen Seminar Essays,* ed. Carlton C. Qualey, 106-30. New York: Columbia University Press, 1968.
———. "New Light on Veblen." In *Essays, Reviews, and Reports,* by Thorstein Veblen. Ed. Joseph Dorfman, 5-326. Clifton, N.J.: Augustus M. Kelley, 1973.
———. "The Source and Impact of Veblen's Thought." In *Thorstein Veblen: A Critical Reappraisal,* ed. Douglas F. Dowd, 1-12. Ithaca: Cornell University Press, 1958.
———. *Thorstein Veblen and His America.* New York: Viking, 1934.
Dos Passos, John. "The Bitter Drink." In *The Big Money.* 1936. Reprinted in *U.S.A.* New York: Modern Library, 1937.
Douglas, Ann. *The Feminization of American Culture.* New York: Avon Press, 1977.
Douglas, Mary. *How Institutions Think.* Syracuse: Syracuse University Press, 1986.
Douglas, Mary, and Baron Isherwood. *The World of Goods: Towards an Anthropology of Consumption.* London: Routledge, 1996.
Dreiser, Helen. *My Life with Dreiser.* Cleveland: World Publishing, 1951.
Dreiser, Theodore. *American Diaries 1902-1926.* Ed. Thomas P. Riggio, James L. W. West III, and Neda M. Westlake. Philadelphia: University of Pennsylvania Press, 1983.

―――. "The American Financier." In *Hey Rub-a-Dub Dub: A Book of the Mystery and Terror and Wonder of Life*, 74-91. New York: Boni and Liveright, 1920.

―――. *An Amateur Laborer.* Ed. Richard W. Dowell, James L. W. West III, and Neda M. Westlake. Philadelphia: University of Pennsylvania Press, 1983.

―――. *An American Tragedy.* 1925. Reprint, New York: Signet, Inc 1964.

―――. *A Book about Myself.* New York: Horace Liveright Inc, 1922.

―――. *The Bulwark.* Garden City: Doubleday, 1946.

―――. "The Country Doctor." In *Twelve Men*, 110-33. New York: Boni and Liveright, 1919.

―――. *Dawn.* New York: Horace Liveright Inc, 1931.

―――. "A Doer of the Word." In *Twelve Men*, 53-75. New York: Boni and Liveright, 1919.

―――. *Dreiser Looks at Russia.* New York: Horace Liveright Inc, 1928.

―――. *The Financier.* New York: Harper and Brothers, 1912.

―――. *The Financier.* 1927. Reprint, New York: Signet, 1967.

―――. *The "Genius."* 1915. Reprint, New York: Boni and Liveright, 1923.

―――. *Hey Rub-a-Dub Dub: A Book of the Mystery and Terror and Wonder of Life.* New York: Boni and Liveright, 1920.

―――. *A Hoosier Holiday.* New York: John Lane Co., 1916.

―――. "I Find the Real American Tragedy." 1935. Reprinted in *Theodore Dreiser: A Selection of Uncollected Prose,* ed. Donald Pizer, 291-99. Detroit: Wayne State University Press, 1977.

―――. *Jennie Gerhardt.* Ed. James L. W. West III. The Pennsylvania Edition. Philadelphia: University of Pennsylvania Press, 1992.

―――. "A Lesson from the Aquarium." Reprinted in *Theodore Dreiser: A Selection of Uncollected Prose,* ed. Donald Pizer, 159-62. Detroit: Wayne State University Press, 1977.

―――. *Letters of Theodore Dreiser.* Ed. Robert H. Elias. 3 vols. Philadelphia: University of Pennsylvania Press, 1959.

―――. "Marriage and Divorce." In *Hey Rub-a-Dub Dub: A Book of the Mystery and Terror and Wonder of Life,* 212-24. New York: Boni and Liveright, 1920.

―――. "A Monarch of Metal Workers." *Success* 2 (June 3, 1899): 453-54.

———. "My Brother Paul." In *Twelve Men*, 76–109. New York: Boni and Liveright, 1919.
———. "Neurotic America and the Sex Impulse." In *Hey Rub-a-Dub Dub: A Book of the Mystery and Terror and Wonder of Life*, 126–41. New York: Boni and Liveright, 1920.
———. *Newspaper Days*. Ed. T. W. Nostwich. The Pennsylvania Edition. Philadelphia: University of Pennsylvania Press, 1991.
———. *Notes on Life*. Ed. Marguerite Tjader and John J. McAleer. University: Alabama University Press, 1974.
———. "Phantom Gold." In *Chains*, 64–97. 1927. Reprint, New York: Howard Fertig, 1987.
———. *Selected Magazine Articles of Theodore Dreiser: Life and Art in the American 1890s*. Ed. Yoshinobu Hakutani. Rutherford: Farleigh Dickinson University Press, 1985.
———. *Sister Carrie*. 1900. Reprint, New York: Signet, 1961.
———. *Sister Carrie*. 1900. Reprint, ed. Donald Pizer. Norton Critical Edition. New York: Norton, 1970.
———. *Sister Carrie*. Ed. John C. Berkey, Alice M. Winters, James L. W. West III, and Neda M. Westlake. The Pennsylvania Edition. New York: Penguin, 1981.
———. *The Stoic*. New York: Doubleday, 1947.
———. *Theodore Dreiser Journalism*. Vol. 1, *Newspaper Writings, 1892–1895*. Ed. T. D. Nostwich. Philadelphia: University of Pennsylvania Press, 1988.
———. *Theodore Dreiser: A Selection of Uncollected Prose*. Ed. Donald Pizer. Detroit: Wayne State University Press, 1977.
———. *Theodore Dreiser's Ev'ry Month*. Ed. Nancy Warner Barrineau. Athens: University of Georgia Press, 1996.
———. *Theodore Dreiser's "Heard in the Corridors" Articles and Related Writings*. Ed. T. D. Nostwich. Ames: Iowa State University Press, 1988.
———. *The Titan*. 1914. Reprint, New York: Signet, 1965.
———. *Tragic America*. New York: Horace Liveright, 1931.
———. "True Art Speaks Plainly." 1903. Reprinted in *Documents of Modern Literary Realism*, ed. George J. Becker, 154–56. Princeton: Princeton University Press, 1963.
———. *Twelve Men*. New York: Boni and Liveright, 1919.
Dreiser, Theodore, and H. L. Mencken. *Dreiser-Mencken Letters: The Correspondence of Theodore Dreiser and H. L. Mencken*. Ed.

Thomas P. Riggio. 2 vols. Philadelphia: University of Pennsylvania Press, 1986.
Dreiser, Vera. *My Uncle Theodore.* New York: Nash, 1976.
Dudley, Dorothy. *Forgotten Frontiers: Dreiser and the Land of the Free.* New York: Harrison Smith and Robert Haas, 1932.
Dyer, Alan W. "Semiotics, Economic Development, and the Deconstruction of the Economic Man." *Journal of Economic Issues* 20:2 (June 1986): 541-49.
———. "Veblen on Scientific Creativity: The Influence of Charles S. Peirce." *Journal of Economic Issues* 20:1 (March 1986): 21-41.
Eastman, Max. *The Literary Mind: Its Place in an Age of Science.* New York: Charles Scribner's Sons, 1935.
Eby, Clare Virginia. "*Babbitt* as Veblenian Critique of Manliness." *American Studies* 34:2 (fall 1993): 5-24.
———. "Cowperwood and Witla, Artists in the Marketplace." *Dreiser Studies* 22:1 (spring 1991): 1-22.
———. "Veblen's Anti-Anti-Feminism." *Canadian Review of American Studies* 1992 Special Issue, Part 2: 215-38.
Edgell, Stephen. "Rescuing Veblen from Valhalla: Deconstruction and Reconstruction of a Sociological Legend." *British Journal of Sociology* 47:4 (December 1996): 627-42.
Eff, E. Anton. "History of Thought as Ceremonial Genealogy: The Neglected Influence of Herbert Spencer on Thorstein Veblen." *Journal of Economic Issues* 23:3 (September 1985): 689-716.
Eley, Geoff. "Nations, Publics, and Political Cultures: Placing Habermas in the Nineteenth Century." Reprinted in *Culture/Power/History*, ed. Nicholas B. Dirks, Geoff Eley, and Sherry B. Ortner, 297-335. Princeton: Princeton University Press, 1994.
Elias, Robert H. *Theodore Dreiser: Apostle of Nature.* Emended ed. Ithaca: Cornell University Press, 1970.
Epstein, Joseph. "A Great Good Girl: Dreiser's 'Jennie Gerhardt.'" *The New Criterion* 2:10 (June 1993): 14-20.
Farrell, James T. "Theodore Dreiser." 1947. Reprinted in *Selected Essays*, ed. Luna Wolf, 150-68. New York: McGraw-Hill, 1964.
Fiedler, Leslie. *Love and Death in the American Novel.* Rev. ed. New York: Stein and Day, 1982.
Fisher, Philip. *Hard Facts: Setting and Form in the American Novel.* New York: Oxford University Press, 1985.
———. "Theodore Dreiser: Promising Dreamers." In *The New Pelican*

Guide to English Literature, ed. Boris Ford. Vol. 9, *American Literature,* 251-62. London: Penguin, 1988.

Fishkin, Shelley Fisher. *From Fact to Fiction: Journalism and Imaginative Writing in America.* Baltimore: Johns Hopkins University Press, 1985.

Fiske, John. "Cultural Studies and the Culture of Everyday Life." In *Cultural Studies,* ed. Lawrence Grossberg, Cary Nelson, and Paul Treicler, 154-64. New York: Routledge, 1992.

Fitzgerald, F. Scott. *The Great Gatsby.* New York: Charles Scribner's Sons, 1925.

Foley, Barbara. *Telling the Truth: The Theory and Practice of Documentary Fiction.* Ithaca: Cornell University Press, 1986.

Foner, Eric. Introduction to *Social Darwinism in American Thought,* by Richard Hofstadter. 1944. Rev. ed. Reprint, Boston: Beacon Press, 1992.

Foucault, Michel. *Discipline and Punish: The Birth of the Prison.* Trans. Alan Sheridan. New York: Vintage Books, 1977.

———. *The History of Sexuality.* Vol. 1, *An Introduction.* Trans. Robert Hurley. New York: Vintage Books, 1990.

Fox, Richard Wightman, and T. J. Jackson Lears, ed. *The Culture of Consumption: Critical Essays in American History,* 1880-1980. New York: Pantheon, 1983.

Furst, Lilian R. *All Is True: The Claims and Strategies of Realist Fiction.* Durham: Duke University Press, 1995.

Fuss, Diana. *Essentially Speaking: Feminism, Nature, and Difference.* New York: Routledge, 1989.

Galbraith, John Kenneth. *The Affluent Society.* 4th ed. New York: Signet, 1984.

———. "A New Theory of Thorstein Veblen." *American Heritage* 24:3 (April 1973): 33-40.

Gammel, Irene. *Sexualizing Power in Naturalism: Theodore Dreiser and Frederick Philip Grove.* Calgary: University of Calgary Press, 1994.

Gates, Henry Louis, Jr. "Good-bye Columbus? Notes on the Culture of Criticism." Reprinted in *The American Literary History Reader,* ed. Gordon Hutner, 245-61. New York: Oxford University Press, 1955.

Gelfant, Blanche H. "What More Can Carrie Want? Naturalistic Ways of Consuming Women." In *The Cambridge Companion to American*

Realism and Naturalism, ed. Donald Pizer, 178-210. Cambridge: Cambridge University Press, 1995.

Gerber, Philip L. "The Financier Himself: Dreiser and C. T. Yerkes." *PMLA* 88:1 (January 1973): 112-21.

———. "Frank Cowperwood: Boy Financier." *Studies in American Fiction* 2:2 (autumn 1974): 165-74.

———. "*Jennie Gerhardt:* A Spencerian Tragedy." In *Dreiser's "Jennie Gerhardt": New Essays on the Restored Text,* ed. James L. W. West III, 77-90. Philadelphia: University of Pennsylvania Press, 1995.

———. *Theodore Dreiser.* New York: Twayne, 1964.

Gilman, Charlotte Perkins. *Women and Economics: A Study of the Economic Relation between Men and Women as a Factor in Social Evolution.* 1898. Reprint, Amherst, N.Y.: Prometheus, 1994.

Gilmore, David D. *Manhood in the Making: Cultural Conceptions of Masculinity.* New Haven: Yale University Press, 1990.

Goldberger, Paul. "The Sameness of Things." *The New York Times Magazine,* April 6, 1997, 56-60.

Gramsci, Antonio. *Prison Notebooks.* Vol. 1. Ed. Joseph A. Buttigieg. Trans. Joseph A. Buttigieg and Antonio Callan. New York: Columbia University Press, 1992.

———. *Selections from the Prison Notebooks by Antonio Gramsci.* Ed. and trans. Quintin Hoare and Geoffrey Nowell Smith. New York: International Publishers, 1971.

Greenwood, Daphne. "The Economic Significance of 'Woman's Place' in Society: A New-Institutionalist View." *Journal of Economic Issues* 13:3 (September 1984): 663-80.

Griffin, Robert. *Thorstein Veblen: Seer of American Socialism.* Hamden, Conn.: The Advocate Press, 1982.

Gunn, Giles. *Thinking across the American Grain: Ideology, Intellect, and the New Pragmatism.* Chicago: University of Chicago Press, 1992.

Habegger, Alfred. *Gender, Fantasy, and Realism in American Literature.* New York: Columbia University Press, 1982.

Hacker, Andrew. "Who They Are." *The New York Times Magazine,* November 19, 1995: 70-71.

Hamilton, David B. "Institutional Economics and Consumption." In *Evolutionary Economics.* Vol. 2, *Institutional Theory and Policy,* ed. Marc R. Tool, 113-36. Armonk, N.Y.: M. E. Sharpe, 1988.

Hapke, Laura. *Tales of the Working Girl: Wage-Earning Women in American Literature, 1890-1925.* New York: Twayne, 1992.

Harper, Phillip Brian. "Fiction and Reform II." In *The Columbia History of the American Novel,* gen. ed. Emory Elliott, 216-39. New York: University of Columbia Press, 1991.

Haskell, Thomas. *The Emergence of Professional Social Science: The American Social Science Association and the Nineteenth-Century Crisis of Authority.* Urbana: University of Illinois Press, 1977.

Haviland, Beverly. "Waste Makes Taste: Thorstein Veblen, Henry James, and the Sense of the Past." *International Journal of Politics, Culture and Society* 7:4 (summer 1994): 615-37.

Heilbroner, Robert. *The Worldly Philosophers.* 3d ed. New York: Simon and Schuster, 1968.

Hochman, Barbara. *The Art of Frank Norris, Storyteller.* Columbia: University of Missouri Press, 1988.

———. "A Portrait of the Artist as a Young Actress: The Rewards of Representation in *Sister Carrie.*" In *New Essays on "Sister Carrie,"* ed. Donald Pizer, 43-65. Cambridge: Cambridge University Press, 1991.

Hodgson, Geoff. "Behind Methodological Individualism." *Cambridge Journal of Economics* 10:3 (September 1986): 211-24.

Hodgson, Geoffrey. *Economics and Evolution: Bringing Life Back into Economics.* Ann Arbor: University of Michigan Press, 1993.

———. "Institutional Economics: Surveying the 'Old' and the 'New.'" Reprinted in *Economics of Institutions,* ed. Geoffrey M. Hodgson, 50-77. Brookfield, Vt.: Edward Elgar, 1993.

Hofstadter, Richard. *Social Darwinism in American Thought.* 1944. Rev. ed. Reprint, Boston: Beacon, 1992.

Hollinger, David A. "The Problem of Pragmatism in American History." In *In the American Province: Studies in the History and Historiography of Ideas,* 23-43. Bloomington: Indiana University Press, 1985.

hooks, bell. "Marginality as Site of Resistance." *Out There: Marginalization and Contemporary Culture,* ed. Russell Ferguson, Martha Gever, Trinh T. Minh-ha, and Cornel West, 341-43. New York: New Museum of Contemporary Art, 1990.

Horwitz, Howard. *By the Law of Nature: Form and Value in Nineteenth Century America.* New York: Oxford University Press, 1991.

Howard, June. *Form and History in American Literary Naturalism.* Chapel Hill: University of North Carolina Press, 1985.

Howells, William Dean. "An Opportunity for American Fiction." 1899. Reprinted in *W. D. Howells as Critic,* ed. Edwin H. Cady, 286–91. London: Routledge and Kegan Paul, 1973.

Humma, John B. "*Jennie Gerhardt* and the Dream of the Pastoral." In *Dreiser's "Jennie Gerhardt": New Essays on the Restored Text,* ed. James L. W. West III, 157–66. Philadelphia: University of Pennsylvania Press, 1995.

Hussman, Lawrence E., Jr. *Dreiser and His Fiction: A Twentieth-Century Quest.* Philadelphia: University of Pennsylvania Press, 1983.

———. "Jennie One-Note: Dreiser's Error in Character Development." In *Dreiser's "Jennie Gerhardt": New Essays on the Restored Text,* ed. James L. W. West III, 43–50. Philadelphia: University of Pennsylvania Press, 1995.

Hutchisson, James M. "The Revision of Theodore Dreiser's *Financier.*" *Journal of Modern Literature* 20:2 (winter 1996): 199–213.

Isernhagen, Hartwig. " 'A Constitutional Inability to Say Yes': Thorstein Veblen, the Reconstruction Program of *The Dial,* and the Development of American Modernism after World War I." *REAL: The Yearbook of Research in English and American Literature* 1 (1982): 153–90.

James, Henry. *The American Scene.* 1907. Reprint, Bloomington: Indiana University Press, 1968.

James, William. *Pragmatism.* 1907. Reprint, Buffalo: Prometheus Books, 1991.

———. *The Principles of Psychology.* 1890. 2 vols. Reprint, New York: Dover, 1950.

———. *Selected Unpublished Correspondence 1885–1910.* Ed. Frederick J. Down Scott. Columbus: Ohio State University Press, 1986.

Jameson, Fredric. *Marxism and Form.* Princeton: Princeton University Press, 1971.

———. *The Political Unconscious: Narrative as a Socially Symbolic Act.* Ithaca: Cornell University Press, 1981.

———. "Reification and Utopia in Mass Culture." *Social Text* 1:1 (winter 1979): 130–48.

Jordan, John M. *Machine-Age Ideology: Social Engineering and American Liberalism, 1911–1939.* Chapel Hill: University of North Carolina Press, 1994.

Josephson, Matthew. *The Robber Barons.* 1934. Reprint, New York: Harcourt, Brace and World, 1962.

Kaplan, Amy. *The Social Construction of American Realism.* Chicago: University of Chicago Press, 1988.

Katope, Christopher G. "*Sister Carrie* and Spencer's *First Principles.*" *American Literature* 41 (1969): 64-75.

Kazin, Alfred. *On Native Grounds: An Interpretation of Modern American Prose.* 1942. Reprint, San Diego: Harcourt Brace Jovanovich, 1982.

Keller, Evelyn Fox. *Reflections on Gender and Science.* New Haven: Yale University Press, 1985.

Kucharski, Judith. "*Jennie Gerhardt:* Naturalism Reconsidered." In *Dreiser's "Jennie Gerhardt": New Essays on the Restored Text,* ed. James L. W. West III, 17-26. Philadelphia: University of Pennsylvania Press, 1995.

Kwiat, Joseph J. "Dreiser's *The 'Genius'* and Everett Shinn, the 'Ash-Can' Painter." *PMLA* 67:2 (March 1952): 15-31.

Lasch, Christopher. *The Culture of Narcissism.* New York: Warner, 1979.

Lears, Jackson. *Fables of Abundance: A Cultural History of Advertising in America.* New York: Basic Books, 1994.

Lears, T. J. Jackson. "Beyond Veblen: Rethinking Consumer Culture in America." In *Consuming Visions: Accumulation and Display of Goods in America, 1880-1920,* ed. Simon J. Bronner, 73-97. New York: Norton, 1989.

———. "From Salvation to Self-Realization: Advertising and the Therapeutic Roots of the Consumer Culture, 1880-1930." In *The Culture of Consumption: Critical Essays in American History, 1880-1980,* ed. Richard Wrightman Fox and T. J. Jackson Lears, 1-38. New York: Pantheon, 1983.

———. *No Place of Grace: Antimodernism and the Transformation of American Culture, 1880-1920.* New York: Pantheon, 1981.

Lehan, Richard. "The City, the Self, and the Modes of Narrative Discourse." In *New Essays on "Sister Carrie,"* ed. Donald Pizer, 65-86. Cambridge: Cambridge University Press, 1991.

———. *Theodore Dreiser: His World and His Novels.* Carbondale: Southern Illinois University Press, 1969.

Lentricchia, Frank. *Ariel and the Police: Michel Foucault, William James, Wallace Stevens.* Madison: University of Wisconsin Press, 1988.

Lepenies, Wolf. *Between Literature and Science: The Rise of Sociology.* Trans. R. J. Hollingdale. Cambridge: Cambridge University Press, 1988.

Lerner, Max. Introduction to *The Portable Veblen,* by Thorstein Veblen, 1-52. New York: Viking, 1958.

———. "Veblen's World." Reprinted in *Ideas Are Weapons: The History and Uses of Ideas,* 138-41. New York: Viking, 1939.

———. "What Is Usable in Veblen?" Reprinted in *Ideas Are Weapons: The History and Uses of Ideas,* 129-38. New York: Viking, 1939.

Levine, George. "By Knowledge Possessed: Darwin, Nature, and Victorian Narrative." *New Literary History* 24 (1993): 363-91.

———. *One Culture: Essays in Science and Literature.* Madison: University of Wisconsin Press, 1987.

———. "Scientific Realism and Literary Representation." *Raritan* 10:1 (spring 1991): 18-39.

Levine, George, ed. *Realism and Representation: Essays on the Problem of Realism in Relation to Science, Literature, and Culture.* Madison: University of Wisconsin Press, 1993.

Lewis, Sinclair. "The American Fear of Literature." 1930. Reprinted in *The Man from Main Street,* by Sinclair Lewis, 3-17. Ed. Harry E. Maule and Melville H. Cane. New York: Pocket Books, 1962.

Liebhafsky, E. E. "The Influence of Charles Saunders Peirce on Institutional Economics." *Journal of Economic Issues* 27:3 (September 1993): 741-54.

Lingeman, Richard. *Theodore Dreiser: An American Journey, 1908-1945.* New York: G. P. Putnam's Sons, 1990.

———. *Theodore Dreiser: At the Gates of the City, 1871-1907.* New York: G. P. Putnam's Sons, 1986.

Livingston, James. "*Sister Carrie*'s Absent Causes." In *Theodore Dreiser: Beyond Naturalism,* ed. Miriam Gogol, 216-46. New York: New York University Press, 1995.

Lukács, Georg. "Narrate or Describe?" In *Writer and Critic and Other Essays,* ed. and trans. Arthur D. Kahn, 110-48. New York: Grosset and Dunlop, 1970.

Lynn, Kenneth. *The Dream of Success: A Study of the Modern American Imagination.* Boston: Little, Brown & Co, 1955.

MacKenna, Stephen. "The Luxury of Lazihead." Reprinted in *Essays Reviews and Reports,* by Thorstein Veblen. Ed. Joseph Dorfman, 615-19. Clifton, N.J.: Augustus M. Kelley, 1973.

Mailer, Norman. *Cannibals and Christians.* New York: Dial Press, 1966.

Marcus, Mordecai. "Loneliness, Death, and Fulfillment in *Jennie Gerhardt*." *Studies in American Fiction* 7 (spring 1979): 61-73.
Martin, Ronald E. *American Literature and the Universe of Force*. Durham, N. C.: Duke University Press, 1981.
Marx, Karl. "Critical Marginal Notes on the Article 'The King of Prussia and Social Reform.' " 1844. Reprinted in *The Marx-Engels Reader*, ed. Robert C. Tucker, 126-32. 2d ed. New York: Norton, 1978.
———. "Economic and Philosophic Manuscripts of 1844." Reprinted in *The Marx-Engels Reader*, ed. Robert C. Tucker, 66-125. 2d ed. New York: Norton, 1978.
———. "Society and Economy in History." In *The Marx-Engels Reader*, ed. Robert C. Tucker, 136-42. 2d ed. New York: Norton, 1978.
Marx, Karl, and Friedrich Engels. *Manifesto of the Communist Party*. 1848. Reprinted in *The Marx-Engels Reader*, ed. Robert C. Tucker, 469-500. 2d ed. New York: Norton, 1978.
Matthiessen, F. O. *Theodore Dreiser*. New York: William Sloane, 1951.
McCloskey, Donald N. *The Rhetoric of Economics*. Madison: University of Wisconsin Press, 1985.
McCracken, Grant. *Culture and Consumption: New Approaches to the Symbolic Character of Consumer Goods and Activities*. Bloomington: Indiana University Press, 1988.
McIlvaine, Robert Morton. "Thorstein Veblen and American Naturalism." Ph.D. diss., Temple University, 1972.
McNamara, Kevin R. "The Ames of the Good Society: *Sister Carrie* and Social Engineering." *Criticism* 34:2 (spring 1992): 217-35.
Mencken, H. L. "Adventures among the New Novels." 1914. Reprinted in *Dreiser-Mencken Letters: The Correspondence of Theodore Dreiser and H. L. Mencken*, ed. Thomas P. Riggio, 2:748-53. Philadelphia: University of Pennsylvania Press, 1986.
———. "Adventures among Books." 1923. Reprinted in *Dreiser-Mencken Letters: The Correspondence of Theodore Dreiser and H. L. Mencken*, ed. Thomas P. Riggio. 2:795-96. Philadelphia: University of Pennsylvania Press, 1986.
———. "The Creed of a Novelist." 1916. Reprinted in *Dreiser-Mencken Letters: The Correspondence of Theodore Dreiser and H. L. Mencken*, ed. Thomas P. Riggio, 2:760-67. Philadelphia: University of Pennsylvania Press, 1986.
———. "The Dreiser Bugaboo." 1917. Reprinted in *Dreiser-Mencken Letters: The Correspondence of Theodore Dreiser and H. L.

Mencken, ed. Thomas P. Riggio, 2:768-75. Philadelphia: University of Pennsylvania Press, 1986.

———. "Dreiser in 840 Pages." 1926. Reprinted in *Dreiser-Mencken Letters: The Correspondence of Theodore Dreiser and H. L. Mencken,* ed. Thomas P. Riggio, 2:796-800. Philadelphia: University of Pennsylvania Press, 1986.

———. "Dreiser's Novel the Story of a Financier Who Loved a Beauty." 1912. Reprinted in *Dreiser-Mencken Letters: The Correspondence of Theodore Dreiser and H. L. Mencken,* ed. Thomas P. Riggio, 2:744-48. Philadelphia: University of Pennsylvania Press, 1986.

———. "A Eulogy for Dreiser." 1947. Reprinted in *Dreiser-Mencken Letters: The Correspondence of Theodore Dreiser and H. L. Mencken,* ed. Thomas P. Riggio, 2:805-6. Philadelphia: University of Pennsylvania Press, 1986.

———. "H. L. Mencken Tells of Dreiser's New Book." 1919. Reprinted in *Dreiser-Mencken Letters: The Correspondence of Theodore Dreiser and H. L. Mencken,* ed. Thomas P. Riggio, 2:790-93. Philadelphia: University of Pennsylvania Press, 1986.

———. "A Novel of the First Rank." 1911. Reprinted in *Dreiser-Mencken Letters: The Correspondence of Theodore Dreiser and H. L. Mencken,* ed. Thomas P. Riggio, 2:740-44. Philadelphia: University of Pennsylvania Press, 1986.

———. "Professor Veblen." Reprinted in *Prejudices: First Series,* 59-82. New York: Alfred A. Knopf, 1919.

———. "Puritanism as a Literary Force." In *A Book of Prefaces,* 197-283. New York: Octagon, 1977.

———. "Theodore Dreiser." 1917. Reprinted in *Dreiser-Mencken Letters: The Correspondence of Theodore Dreiser and H. L. Mencken,* ed. Thomas P. Riggio, 2:775-90. Philadelphia: University of Pennsylvania Press, 1986.

Merton, Robert K. *Social Theory and Social Structure.* Enlarged ed. New York: The Free Press, 1968.

Metzger, Walter P. *Academic Freedom in the Age of the University.* New York: Columbia University Press, 1955.

Michaels, Walter Benn. *The Gold Standard and the Logic of Naturalism: American Literature at the Turn of the Century.* Berkeley: University of California Press, 1987.

Miller, Edythe S. "Veblen and Women's Lib.: A Parallel." *Journal of Economic Issues* 6:2, 3 (September 1972): 75-86.

Mills, C. Wright. Introduction to the Mentor Edition of *The Theory of the Leisure Class,* by Thorstein Veblen. New York: Mentor, 1953.
———. *The Sociological Imagination.* New York: Oxford University Press, 1959.
Mitchell, Lee Clark. *Determined Fictions: American Literary Naturalism.* New York: Columbia University Press, 1989.
Mitchell, W. C. "Thorstein Veblen: 1857-1929." Reprinted in *Essays, Reviews, and Reports,* by Thorstein Veblen. Ed. Joseph Dorfman, 601-6. Clifton, N.J.: Augustus M. Kelley, 1973.
Mizruchi, Susan. "Fiction and the Science of Society." In *The Columbia History of the American Novel,* gen. ed. Emory Elliott, 189-215. New York: Columbia University Press, 1991.
———. *The Power of Historical Knowledge: Narrating the Past in Hawthorne, James, and Dreiser.* Princeton: Princeton University Press, 1988.
Moers, Ellen. *Two Dreisers.* New York: Viking, 1969.
Mukherjee, Arun. *The Gospel of Wealth in the American Novel: The Rhetoric of Dreiser and Some of His Contemporaries.* London: Croom Helm, 1987.
Mulvey, Laura. *Visual and Other Pleasures.* Bloomington: Indiana University Press, 1989.
Myers, Gustavus. *History of the Great American Fortunes.* 1910. New York: Random House, 1936.
Noble, David. "Dreiser and Veblen and the Literature of Cultural Change." In *Studies in American Culture: Dominant Ideas and Images,* ed. Joseph J. Kwiat and Mary C. Turpie, 139-52. Minneapolis: University of Minnesota Press, 1966.
———. "The Sacred and the Profane: The Theology of Thorstein Veblen." In *Thorstein Veblen: The Carleton College Veblen Seminar Essays,* ed. Carlton C. Qualey, 72-105. New York: Columbia, 1968.
Norris, Frank. *McTeague.* 1899. Reprint, New York: Penguin, 1982.
———. *The Octopus.* 1901. Reprint, New York: Signet, 1964.
———. *The Pit.* 1903. Reprint, New York: Grove Press, 1956.
———. *The Responsibilities of the Novelist.* Reprinted with *Criticism and Fiction,* by William Dean Howells. New York: Hill and Wang, 1962.
Novick, Peter. *That Noble Dream: The "Objectivity Question" and the American Historical Profession.* Cambridge: Cambridge University Press, 1988.

Orlov, Paul. "Technique as Theme in *An American Tragedy.*" Reprinted in *Theodore Dreiser's "An American Tragedy,"* ed. Harold Bloom, 85-102. New York: Chelsea House Publishers, 1988.

Orvell, Miles. *The Real Thing: Imitation and Authenticity in American Culture, 1880-1940.* Chapel Hill: University of North Carolina Press, 1989.

Oxford English Dictionary. Oxford: Oxford University Press, 1980.

Parrington, Vernon Louis. *Main Currents in American Thought.* Vol. 3, *The Beginnings of Critical Realism in America: 1860-1920.* New York: Harcourt, Brace, and World, 1930.

Peirce, Charles S. "The Scientific Attitude and Fallibilism." Reprinted in *Philosophical Writings of Peirce,* ed. Justus Buchler, 42-59. New York: Dover, 1955.

Petrey, Sandy. "The Language of Realism, the Language of False Consciousness: A Reading of *Sister Carrie.*" *Novel* 10:2 (winter 1977): 101-13.

Peyser, Thomas Galt. "Reproducing Utopia: Charlotte Perkins Gilman and *Herland.*" *Studies in American Fiction* 20:1 (1992): 1-16.

Pizer, Donald. "Dreiser and the Naturalistic Drama of Consciousness." *Journal of Narrative Technique* 21:2 (spring 1991): 202-11.

———. Introduction to *New Essays on* Sister Carrie, 1-22. Cambridge: Cambridge University Press, 1991.

———. *The Novels of Theodore Dreiser: A Critical Study.* Minneapolis: University of Minnesota Press, 1976.

Pizer, Donald, ed. *New Essays on "Sister Carrie."* Cambridge: Cambridge University Press, 1991.

Posnock, Ross. "Henry James, Veblen, and Adorno: The Crisis of the Modern Self." *Journal of American Studies* 21:1 (April 1987) 31-54.

"A Protest against the Suppression of Theodore Dreiser's *The 'Genius.'*" Reprinted in *Dreiser-Mencken Letters: The Correspondence of Theodore Dreiser and H. L. Mencken,* ed. Thomas P. Riggio, 2: 802-4. Philadelphia: University of Pennsylvania Press, 1986.

Qualey, Carlton C. Introduction to *Thorstein Veblen: The Carleton College Veblen Seminar Essays,* 1-15. New York: Columbia University Press, 1968.

Rabine, Leslie W. "A Woman's Two Bodies: Fashion Magazines, Consumerism, and Feminism." In *On Fashion,* ed. Shari Benstock and Suzanne Ferriss, 59-75. New York: Rutgers University Press, 1994.

Rahv, Philip. "Notes on the Decline of Naturalism." Reprinted in *Documents of Modern Literary Realism,* ed. George J. Becker, 579-90. Princeton: Princeton University Press, 1953.

Riesman, David. *Thorstein Veblen: A Critical Interpretation.* New York: Charles Scribner's Sons, 1953.

Riggio, Thomas P. "Carrie's Blues." In *New Essays on "Sister Carrie,"* ed. Donald Pizer, 23-42. Cambridge: Cambridge University Press, 1991.

———. "Theodore Dreiser: Hidden Ethnic." *MELUS* 11:1 (spring 1984): 53-63.

Roberts, Sidney I. "Portrait of a Robber Baron: Charles T. Yerkes." *Business History Review* 35 (1961): 341-71.

Rockefeller, John D. *Random Reminiscences of Men and Events.* New York: Doubleday, Doran, 1937.

Rojek, Chris. "Baudrillard and Leisure." *Leisure Studies* 9 (January 1990): 7-20.

Romero, Lora. "Domesticity and Fiction." In *The Columbia History of the American Novel,* gen. ed. Emory Elliott, 110-29. New York: Columbia University Press, 1991.

———. "Bio-Political Resistance in Domestic Ideology and *Uncle Tom's Cabin.*" Reprinted in *The American Literary History Reader,* ed. Gordon Hutner, 111-30. New York: Oxford University Press, 1995.

Rorty, Richard. *Consequences of Pragmatism.* Minneapolis: University of Minnesota Press, 1982.

———. *Philosophy and the Mirror of Nature.* Princeton: Princeton University Press, 1979.

Ross, Dorothy. *The Origins of American Social Science.* Cambridge: Cambridge University Press, 1991.

Ross, Dorothy, ed. *Modernist Impulses in the Human Sciences 1870-1930.* Baltimore: Johns Hopkins University Press, 1994.

Ross, Valerie. "Chill History and Rueful Sentiments in *Jennie Gerhardt.*" In *Dreiser's "Jennie Gerhardt": New Essays on the Restored Text,* ed. James L. W. West III, 27-42. Philadelphia: University of Pennsylvania Press, 1995.

Rubin, Gayle. "Thinking Sex: Notes for a Radical Theory of the Politics of Sexuality." Reprinted in *The Lesbian and Gay Studies Reader,* ed. Henry Abelove, Michele Aina Barale, and David M. Halperin, 3-44. New York: Routledge, 1993.

Said, Edward W. *Representations of the Intellectual.* New York: Vintage, 1996.

Salzman, Jack. "The Publication of *Sister Carrie:* Fact and Fiction." *Library Chronicle* 33 (1967): 119-33.

Samuels, Warren J. Introduction to *The Place of Science in Modern Civilization and Other Essays,* by Thorstein Veblen, vii-xxx. Reprint, New Brunswick, N.J.: Transaction Publishers, 1990.

———. "The Self-Referentiability of Thorstein Veblen's Theory of the Preconceptions of Economic Science." *Journal of Economic Issues* 24:3 (September 1990): 695-718.

Scharnhorst, Gary. "Reconstructing Here Also: On the Later Poetry of Charlotte Perkins Gilman." In *Critical Essays on Charlotte Perkins Gilman,* ed. Joanne B. Karpinski, 249-68. New York: G. K. Hall, 1992.

Schimmer, Ralf. *Populismus und Sozialwissenschaften im Amerika der Jahrhundertwende.* Frankfurt: Campus, 1996.

———. "Wider die Legende von der unüberbrückbaren Distanz: Der amerikanische Populismus als normativer Grundgehalt der Veblenschen Sozialkritik." In *Ziet der Institutionen. Thorstein Veblens evolutorische Ökonomik,* ed. Reinhard Penz and Holger Willcop. Marburg: Metropolis, 1996.

Schneider, Louis. *The Freudian Psychology and Veblen's Social Theory.* New York: King's Crown Press, 1948.

Schöpp, Joseph C. "Cowperwood's Will to Power: Dreiser's *Trilogy of Desire* in the Light of Nietzsche." In *Nietzsche in American Literature and Thought,* ed. Manfred Pütz, 139-54. Columbia, S. C.: Camden House, 1995.

Schorer, Mark. *Sinclair Lewis: An American Life.* New York: McGraw-Hill, 1961.

Schwartz, Carol A. "*Jennie Gerhardt:* Fairy Tale as Social Criticism." *American Literary Realism* 19:2 (winter 1987): 16-29.

See, Fred G. *Desire and the Sign: Nineteenth-Century American Fiction.* Baton Rouge: Louisiana State University Press, 1987.

Seltzer, Mark. *Bodies and Machines.* New York: Routledge, 1992.

Shannon, Christopher. *Conspicuous Criticism: Tradition, the Individual, and Culture in American Social Thought, from Veblen to Mills.* Baltimore: Johns Hopkins University Press, 1996.

Shi, David. E. *Facing Facts: Realism in American Thought and Culture, 1850-1920.* New York: Oxford University Press, 1995.

Shloss, Carol. *In Visible Light: Photography and the American Writer, 1840–1940.* New York: Oxford University Press, 1987.

Silverman, Kaja. "Fragments of a Fashionable Discourse." In *On Fashion,* ed. Shari Benstock and Suzanne Ferriss, 183–96. New York: Rutgers University Press, 1994.

Simmel, George. "Fashion." *International Quarterly* 10 (October 1904): 130–55.

Slater, Gilbert. "The Psychological Basis of Economic Theory, Part 2." *Sociological Review* 15 (1923): 278–85.

Smith, Carl. *Chicago and the American Literary Imagination, 1880–1920.* Chicago: Chicago University Press, 1984.

Sollors, Werner. *Beyond Ethnicity: Consent and Descent in American Culture.* New York: Oxford University Press, 1986.

Solomon-Godeau, Abigail. "The Other Side of Venus: The Visual Economy of Feminine Display." In *The Sex of Things: Gender and Consumption in Historical Perspective,* ed. Victoria De Grazia and Ellen Furlough, 113–49. Berkeley: University of California Press, 1996.

Spencer, Herbert. *First Principles.* 1862. Reprint, New York: The De Witt Revolving Fund, Inc., 1958.

Spindler, Michael. *American Literature and Social Change: William Dean Howells to Arthur Miller.* London: Macmillan, 1983.

Strychacz, Thomas. *Modernism, Mass Culture, and Professionalism.* Cambridge: Cambridge University Press, 1993.

Suleiman, Susan Rubin. *Authoritarian Fictions: The Ideological Novel as a Literary Genre.* New York: Columbia University Press, 1983.

Sumner, William Graham. *What Social Classes Owe to Each Other.* 1883. Reprint, Caldwell, Idaho: The Caxton Printers, 1963.

Sundquist, Eric J. "The Country of the Blue." In *American Realism: New Essays,* 3–24. Baltimore: Johns Hopkins University Press, 1982.

Susman, Warren I. " 'Personality' and the Making of Twentieth-Century Culture." In *Culture as History: The Transformation of American Society in the Twentieth Century,* 271–85. New York: Pantheon, 1973.

Swanberg, W. A. *Dreiser.* New York: Charles Scribner's Sons, 1965.

Tarbell, Ida M. *The History of the Standard Oil Company.* 1904. Briefer version, ed. David M. Chalmers. New York: Norton, 1969.

"Theodore Dreiser Now Turns to High Finance." 1912. Reprinted in

Theodore Dreiser: A Selection of Uncollected Prose, ed. Donald Pizer, 196-99. Detroit: Wayne State University Press, 1977.

Thomas, Brook. *American Literary Realism and the Failed Promise of Contract.* Berkeley: University of California Press, 1997.

Tilman, Rick. *The Intellectual Legacy of Thorstein Veblen: Unresolved Issues.* Westport, Conn.: Greenwood Press, 1996.

———. *Thorstein Veblen and His Critics 1891-1963: Conservative, Liberal, and Radical Perspectives.* Princeton: Princeton University Press, 1992.

Tompkins, Jane. *Sensational Designs: The Cultural Work of American Fiction 1790-1860.* New York: Oxford University Press, 1985.

Toulouse, Teresa. "Veblen and His Reader: Rhetoric and Intention in *The Theory of the Leisure Class.*" *The Centennial Review* 29:2 (spring 1985): 249-67.

Trachtenberg, Alan. *The Incorporation of America: Culture and Society in the Gilded Age.* New York: Hill and Wang, 1982.

———. "Who Narrates? Dreiser's Presence in *Sister Carrie.*" In *New Essays on "Sister Carrie,"* ed. Donald Pizer, 87-122. Cambridge: Cambridge University Press, 1991.

Trilling, Lionel. *The Liberal Imagination.* 1950. Reprint, New York: Viking Press, 1953.

Trump, Donald J., with Tony Schwartz. *The Art of the Deal.* New York: Random House, 1987.

Valverde, Mariana. "The Love of Finery: Fashion and the Fallen Woman in Nineteenth-Century Social Discourse." *Victorian Studies* 32:2 (winter 1989): 169-88.

Van Sickle, Larry. "The Pathologizing of Thorstein Veblen: He Ain't No Lord Keynes but He Just Might Be Redemptive." Presented at Western Social Science Association, April 23-29, 1997. Albuquerque, NM.

Veblen, Thorstein. *Absentee Ownership and Business Enterprise in Recent Times: The Case of America.* 1923. Reprint, Boston: Beacon Press, 1967.

———. "Christian Morals and the Competitive System." 1910. Reprinted in *Essays in Our Changing Order,* ed. Leon Ardzrooni, 200-218. New York: Augustus M. Kelley, 1964.

———. "Dementia Praecox." 1922. Reprinted in *Essays in Our Changing Order,* ed. Leon Ardzrooni, 423-36. New York: Augustus M. Kelley, 1964.

———. "The Economic Theory of Women's Dress." 1894. Reprinted in *Essays in Our Changing Order*, ed. Leon Ardzrooni, 65-77. New York: Augustus M. Kelley, 1964.

———. *The Engineers and the Price System.* 1921. Reprint, New York: Harcourt, Brace and World, 1963.

———. *Essays in Our Changing Order.* Ed. Leon Ardzrooni. 1934. Reprint, New York: Augustus M. Kelley, 1964.

———. *Essays, Reviews, and Reports.* Ed. Joseph Dorfman. Clifton, N.J.: Augustus M. Kelley, 1973.

———. "The Evolution of the Scientific Point of View." 1908. Reprinted in *The Place of Science in Modern Civilization*, 32-55. New Brunswick, N.J.: Transaction Publishers, 1990.

———. *The Higher Learning in America: A Memorandum on the Conduct of Universities by Business Men.* 1918. Reprint, Stanford: Academic Reprints, 1954.

———. *Imperial Germany and the Industrial Revolution.* New York: Macmillan, 1915.

———. *An Inquiry into the Nature of Peace and the Terms of Its Perpetuation.* New York: Macmillan, 1917.

———. "The Instinct of Workmanship and the Irksomeness of Labor." 1898. Reprinted in *Essays in Our Changing Order*, ed. Leon Ardzrooni, 78-96. New York: Augustus M. Kelley, 1964.

———. *The Instinct of Workmanship and the State of the Industrial Arts.* 1914. New York: Norton, 1941.

———. "The Intellectual Preeminence of Jews in Modern Europe." 1919. Reprinted in *Essays in Our Changing Order*, ed. Leon Ardzrooni, 219-31. New York: Augustus M. Kelley, 1964.

———. "Menial Servants During the Period of the War." 1918. Reprinted in *Essays in Our Changing Order*, ed. Leon Ardzrooni, 267-78. New York: Augustus M. Kelley, 1964.

———. "Mr. Cummings's Strictures on 'The Theory of the Leisure Class.'" 1899. Reprinted in *Essays in Our Changing Order*, ed. Leon Ardzrooni, 16-31. New York: Augustus M. Kelley, 1964.

———. "The Mutation Theory and the Blond Race." Reprinted in *The Place of Science in Modern Civilization*, 457-76. New Brunswick, N.J.: Transaction Publishers, 1990.

———. "The Opportunity of Japan." 1915. Reprinted in *Essays in Our Changing Order*, ed. Leon Ardzrooni, 248-66. New York: Augustus M. Kelley, 1964.

———. *The Place of Science in Modern Civilization and Other Essays.* 1919. Reprint, New Brunswick, N.J.: Transaction Publishers, 1990.

———. *The Portable Veblen,* ed. Max Lerner. New York: Viking Press, 1958.

———. "The Socialist Economics of Karl Marx and His Followers: I." 1906. Reprinted in *The Place of Science in Modern Civilization,* 409-30. New Brunswick, N.J.: Transaction Publishers, 1990.

———. "Some Neglected Points in the Theory of Socialism." 1908. Reprinted in *The Place of Science in Modern Civilization,* 387-408. New Brunswick, N.J.: Transaction Publishers, 1990.

———. *The Theory of Business Enterprise.* 1904. Reprint, New York: Charles Scribner's Sons, 1919.

———. *The Theory of the Leisure Class: An Economic Study of Institutions.* 1899. Reprint, New York: Modern Library, 1931.

———. *The Vested Interests and the Common Man.* 1919. Reprint, New York: Viking, 1933.

———, trans. *The Laxdæla Saga.* New York: B. W. Huebsch, 1925.

Vidich, Arthur J. "Veblen and the PostKeynesian Political Economy." *International Review of Sociology* (Rome) 3 (1992): 151-81.

Vidich, Arthur J., and Joseph Bensman. *Small Town in Mass Society: Class, Power and Religion in a Rural Community.* Rev. ed. Princeton: Princeton University Press, 1968.

Vivas, Eliseo. "Dreiser, an Inconsistent Mechanist." Reprinted in *Critical Essays on Theodore Dreiser,* ed. Donald Pizer, 30-37. Boston: G. K. Hall & Co., 1981.

Waddoups, Jeffrey, and Rick Tilman. "Thorstein Veblen and the Feminism of Institutional Economics." *International Review of Sociology* (Rome) 3 (1992): 182-204.

Wadlington, Warwick. "Pathos and Dreiser." Reprinted in *Critical Essays on Theodore Dreiser,* ed. Donald Pizer, 213-27. Boston: G. K. Hall & Co., 1981.

Walcutt, Charles Child. *American Literary Naturalism: A Divided Stream.* Minneapolis: University of Minnesota Press, 1956.

Waller, William T., Jr. "The Concept of Habit in Economic Analysis." Reprinted in *Economics of Institutions,* ed. Geoffrey M. Hodgson, 9-22. Brookfield, Vt.: Edward Elgar, 1993.

Waller, William, and Linda R. Robertson. "Why Johnny (Ph.D., Economics) Can't Read: A Rhetorical Analysis of Thorstein Veblen and a Response to Donald McCloskey's *Rhetoric of Economics."* *Journal of Economic Issues* 24:4 (December 1990): 1027-44.

Ward, Lester Frank. "The Theory of the Leisure Class." 1900. Reprinted in *Essays, Reviews, and Reports,* by Thorstein Veblen. Ed. Joseph Dorfman, 619-30. Clifton, N.J.: Augustus M. Kelley, 1973.

———. *Pure Sociology: A Treatise on the Origin and Spontaneous Development of Society.* 2d. ed. New York: Macmillan, 1907.

Warren, Robert Penn. *Homage to Theodore Dreiser.* New York: Random House, 1971.

Weber, Max. *The Protestant Ethic and the Spirit of Capitalism.* 1905. Trans. Talcott Parsons. Reprint, London: George Allen and Unwin Ltd., 1930.

West, Cornel. "The Postmodern Crisis of the Black Intellectuals." In *Cultural Studies,* ed. Lawrence Grossberg, Cary Nelson, and Paul Treicler, 689-96. New York: Routledge, 1992.

West, James L. W., III, ed. *Dreiser's "Jennie Gerhardt": New Essays on the Restored Text.* Philadelphia: University of Pennsylvania Press, 1995.

Wharton, Edith. *The Age of Innocence.* 1920. Reprint, New York: Collier Macmillan, 1992.

White, Morton. *Social Thought in America: The Revolt against Formalism.* Boston: Beacon Press, 1947.

Wiebe, Robert H. *Businessmen and Reform: A Study of the Progressive Movement.* Cambridge: Harvard University Press, 1962.

Williams, Raymond. *Marxism and Literature.* Oxford: Oxford University Press, 1977.

Wilson, Christopher P. *The Labor of Words: Literary Professionalism in the Progressive Era.* Athens: University of Georgia Press, 1985.

———. "Labor and Capital in *Jennie Gerhardt.*" In *Dreiser's "Jennie Gerhardt": New Essays on the Restored Text,* ed. James L. W. West III, 103-14. Philadelphia: University of Pennsylvania Press, 1995.

———. "*Sister Carrie* Again." *American Literature* 53:2 (May 1981): 287-90.

Wilson, Daniel J. *Science, Community, and the Transformation of American Philosophy, 1860-1930.* Chicago: University of Chicago Press, 1990.

Wolstenholme, Susan. "Brother Theodore, Hell on Women." In *American Novelists Revisited: Essays in Feminist Criticism,* ed. Fritz Fleischmann, 243-64. Boston: G. K. Hall, 1982.

Woolf, Virginia. *A Room of One's Own.* 1929. Reprint, New York: Harvest/HBJ, 1957.

Yeazell, Ruth Bernard. "The Conspicuous Wasting of Lily Bart." *ELH* 59:3 (fall 1992): 713-34.

Zanine, Louis J. *Mechanism and Mysticism: The Influence of Science on the Thought and Work of Theodore Dreiser.* Philadelphia: University of Pennsylvania Press, 1993.

Ziff, Larzer. *The American 1890s: Life and Times of a Lost Generation.* New York: Viking, 1966.

Zola, Emile. "The Experimental Novel." 1880. Reprinted in *Documents of Modern Literary Realism,* ed. George J. Becker, 161-96. Princeton: Princeton University Press, 1963.

INDEX

Absentee ownership, 77n13
Adorno, T. W. 152, 161
Agnew, Jean-Christophe, 119n18
Albany Times-Union, 135
Alden, Roberta, 5, 10, 45, 64, 139-88 *passim*
American Association of University Professors, 50
American culture: social values of, 149, 150, 178-79
American Match Company, 73, 74
Ames, Bob, 78, 126, 127
Anderson, Maxwell, 7
Anderson, Rasmus Bjorn, 15n26
Anti-intellectualism: in American culture, 58-59, 62
Archer, Newland, 176
Arendt, Hannah, 158
Art Institute of Chicago, 26
Ash-Can painting, 25, 26
Ayres, C. E., 159, 162

Bakhtin, M. M., 14
Baldwin, James, 17, 184
Banta, Martha, 182n41
Barrineau, Nancy Warner, 81, 153
Barton, Bruce, 86; *The Man Nobody Knows,* 38
Baudrillard, Jean, 13
Baym, Nina, 180, 181
Bell, Michael Davitt, 183
Bellamy, Edward: *Looking Backward,* 15
Bellow, Saul, 184

Bensman, Joseph, 54n41
Bentham, Jeremy, 30n14, 119
Berger, John, 122, 123n26
Bersani, Leo, 9, 127n32
Berthoff, Warner, 77, 152-53
Blue, Angela, 27, 28, 29
Booth, Franklin, 61
Bouguereau, 26, 27
Bowlby, Rachel, 109n5
Brander, Senator, 153, 167, 168, 169, 170, 174, 177, 183
Brennan, Stephen C., 60, 83
Briggs, Hortense, 140
British East India Co., 70
Brown, Grace "Billie," 135
Bryan, William Jennings, 73
Burgess, Ernest, 5
Business: and ceremonialism, 96; and creativity, 68-69, 78-79, 86-87, 88, 89, 90, 91-92, 103-4, 106; evolution of, 91; immaterial basis of, 97-103, 106, 107; and politics, 105; role of, 18, 67; romanticizing of, 91; and sentimentalism, 187; and status quo, 188; tycoons as moral revolutionaries, 84-85, 88-93, 95-96, 103, 106; tycoons as predators, 68, 72, 74-75, 76-78, 88-89, 96-97, 98, 99, 102, 103, 105, 106, 112-13, 137, 171-73, 178-79; versus industry, 66, 68, 69-72, 75-78, 79, 91, 97, 98, 106
Butler, Edward Malia, 70, 72, 83, 92
Byrne, David, 146

219

Calvinism, 75
Canada, 111
Capitalism, 9-10, 18, 19, 25, 53-54; as a calling, 75-76, 80; evolution of, 81-82, 83; impact of, 24, 25; as predatory system, 97, 99, 102-3, 105, 106, 108-13, 141, 172-73; role of psychology in, 107-14; role of superstition in, 66; seductiveness of, 67, 118, 128; and status quo, 82; versus communalism, 70-71
Cappetti, Carla, 6
Captains of Industry, 69, 105
Captains of Solvency, 69, 72, 79, 86
Carleton College, 36
Carnegie, Andrew, 79, 100
Case studies: use of, 16, 33
Catholic Church, 40, 41. *See also* Religion
Censorship, 34; and suppression of *Sister Carrie*, 60-61
Ceremonialism, 12-14, 159, 162; in business, 96; and class status, 138-40; and marriage, 170; versus truth, 145-46
Chase, Richard, 114, 181
Chicago, Ill., 59, 72, 78, 79, 83-84, 91, 99, 100, 102, 124n27, 119, 126, 131, 133; tunnels of, 77
Chicago fire, 72, 74, 92
Chicago "quadrumvirate," 73, 74
Chicago school (sociology), 5
Christianity. *See* Catholic Church; Religion
Cincinnati, Ohio, 169
Citizens' Committee of One Hundred, 84
Civil War, 52, 75, 76, 99n43
Clark, John Bates, 36, 37
Clark, John Maurice, 35n23
Class status: and fashion, 122-23, 124-26; and identity, 64, 118-19, 122-23, 130-47; obsession with, 3, 118-19, 122-23, 130-47 *passim*; and resistance to change, 176; and sexuality, 168; and women's work, 165-67; and work, 167-68
Class system, 2; American enthusiasm for, 5; as predatory, 161
Cleveland, Ohio, 169

Committee on Public Information, 14
Commons, John, 2n2
Comte, Auguste, 163
Communism, 44, 87
Comstock, Anthony, 56
Comstock Act, 57n43
Conspicuous consumption, 1, 107-9; by women, 163, 169, 174
Consumer culture, 13, 107-10; and sexuality, 118
Convention: critical war on, 63-64. *See also* Institutionalism
Cooke, Jay, 75, 78, 79
Cowley, Malcolm, 35n23
Cowperwood, Aileen, 78, 83, 91, 92
Cowperwood, Berenice Fleming, 98n42, 104
Cowperwood, Frank, 5, 18, 64-106 *passim*, 150, 151, 156, 161, 172, 175, 188, 189
Cowperwood, Lillian, 63, 83,
Cowperwood and Company, 97
Creation, 83
Crime: American obsession with, 10-11, 135, 143, 144-46
Cultural criticism, ix-xii, 2, 10-11, 17, 18, 19, 59, 83, 84, 89, 148-49, 172; history of, 31; interdisciplinary approach to, 3-4, 5-6, 12, 16; and literature, 5-6; and sentimentalism, 179-88; use of science in, 43
Cummings, John, 15-16, 24

Dale, Suzanne, 174
Darwin, Charles, 27, 28, 29, 83n24, 128n33; *The Descent of Man*, 164; *Origin of Species*, 30, 94
Daumier, 104
Degler, Carl, 27
De Grazia, Victoria, 108-9n3
Deleuze, Gilles, x
Denver, Colo., 144
Deshon, Florence, 7
Dewey, John, x, 30
Dial, the, 7, 16, 113
Diggins, John, 21
Dorfman, Joseph, 7, 13n22, 47, 48-49, 66, 162, 185

Dos Passos, John, 13
Doubleday, 61
Doubleday, Mrs. Frank, 60
Double voicedness, 14
Douglas, Mary, 20, 35
Dreiser, Ed (brother), 134
Dreiser, Emma (sister), 183
Dreiser, Helen (second wife), 90n34
Dreiser, John Paul (father), 40
Dreiser, John Paul, Jr. (brother Paul Dresser), 182-84
Dreiser, Sarah (mother), 154, 183n45
Dreiser, Theodore
—Life: career as journalist, 59-60; career as novelist, 60-62; and censorship, 61; as critic, ix, x, 2, 8-9, 10-11; education of, 8; emotional nature of, 183-84; family of, 7, 8, 40, 90, 134, 154, 182-84; heroes of, 58; influence of Veblen on, 6-7, 8, 137; memories, use of, 59, 157-58; as outsider, 10, 17, 21, 22, 46-48, 56-58, 149; projection of self, 66-67; and religion, 93-96; reputation as womanizer, 10; response to critics, 24; self-construction of, 146-47; sexuality, importance of, 41-42, 151, 183-84; style of, x-xii, 11, 12, 18, 21-22, 29, 33, 43; talent of, as "feminine," 183; values of, 149, 150, 165; view on sentimentalism, 182-85; view on work, 157-58; works, reappraisal of, x, xi, 181-82, 183-85
—Works: *An Amateur Laborer,* 125-26, 157, 158; "The American Financier," 68-69, 84; *An American Tragedy,* xi, 5, 7, 11, 13, 18, 33, 43, 45, 113, 117, 135-157 *passim,* 171, 173; *A Book about Myself,* 21, 22, 59, 60, 146; *The Bulwark,* 40; *Dawn,* 8, 11, 21, 40-41, 93, 124n28, 128, 154; *Dreiser Looks at Russia,* 8, 9, 44, 57, 86; *The Financier,* 5, 33, 66-107 *passim; A Gallery of Women,* 7, 42; *The "Genius,"* 5, 17, 21, 25-29, 42, 66, 157; *Hey Rub-a-Dub-Dub,* 8, 21; *A Hoosier Holiday,* 8, 40, 58, 61, 78; "I Find the Real American Tragedy," 137; *Jennie Gerhardt,* 5, 18, 40, 149-89 *passim;* "My Brother Paul," 182, 183, 184; "Neurotic America and the Sex Impulse," 33, 34, 35, 36; *Newspaper Days,* 6, 41, 134; *Notes on Life,* 116; "On the Banks of the Wabash," 184; "Phantom Gold," 87n30; "The Professional Intellectual and His Present Place," 11; *Sister Carrie,* xi, 1, 5, 6, 13, 18, 22, 33, 60, 61, 76-179 *passim; The Stoic,* 5, 66, 76-102 *passim; The Titan,* 5, 66, 76-102 *passim,* 171; *Tragic America,* 8; *The Trilogy of Desire,* 5, 17-18, 64-106 *passim,* 148; "True Art Speaks Plainly," 23, 24; *Twelve Men,* 7, 157, 174, 182
Dreiser, Vera (niece), 183n45
Dresser, Paul. *See* Dreiser, John Paul, Jr.
Drouet, Charles, 86, 113, 118, 122, 123, 124, 126, 129, 165n23
Dudley, Dorothy, 45, 57n44, 60, 63
Dyer, Alan W., 51

Eastman, Max: *The Literary Mind: Its Place in an Age of Science,* 33
Economic Man, 109, 120
Economics, 2, 3; and American culture, 87; and anthropology, 159-60; evolution of theory, 119-21; and literature, 65-66; maxims of, 30-31; modernism in, 15-16; as a science, 113-14; theories of, 67, 96, 105; theory of, and religion, 37-38, 39
Economy: manipulation of, 73-75, 76, 85, 87, 88, 89, 96-103, 105
Ellis, Henry Havelock, 34
Emerson, Ralph Waldo, 4
Epstein, Joseph, 181
Evolution: of business, 91; of capitalist culture, 81-82, 83; of economic theory, 119-21; of humans, 93-94, 163-64. *See also* Darwin, Charles; Spencer, Herbert
Ev'ry Month (magazine), 81

Farrell, James T., 135-36
Fashion: and class status, 132, 134, 135; and identity, 125-26
Feminism, 35, 59; and cultural criticism, 47; and fashion, 124-25

222 INDEX

Fiedler, Leslie, 175, 179, 182
Financiers. *See* Business: tycoons
Finchleys, 138
Fisher, Philip, 10, 114
Fitzgerald, F. Scott: *The Great Gatsby,* 91
Fitzgerald, Zelda, 184n46
Fitzgerald and Moy's, 122, 133
Fleming, Berenice, 90, 174
Foner, Eric, 27
Foucault, Michel, x, xi, 11, 42, 43, 119 168n26
Four Hundred, the, 137
Freud, Sigmund, 8, 34, 141, 159
Furst, Lilian R., 32n19
Fuss, Diana, 43n31, 152

Galbraith, John Kenneth, 7, 110n6
Gammel, Irene, 42
Gap (store), 108
Gates, Henry Louis, 47n33
Gatsby, Jay, 91
Gender, 132; anxiety among men, 173-75; essentialism, 148-89 *passim*. *See also* Feminism; Masculinity; Women
Gerald, Letty Pace, 176, 178
Gerber, Philip L., 66-67n3, 80-81, 104n47
Gerhardt, Jennie, 42, 64, 149-89 *passim*
Gerhardt, Mr., 174
Gillette, Chester, 135
Gilman, Charlotte Perkins, 27, 28n13; "The Oyster and the Starfish," 93n36; *Women in Economics,* 165
Gilmore, David D., 175
Goldberger, Paul, 108
Gold standard, 73n10
Gould, Jay, 84
Gramsci, Antonio, 112
Grand Rapids, Ohio, 146
Green-Davidson Hotel, 113, 140, 142
Griffiths, Clyde, 5, 10, 43, 45, 64, 86, 107, 113-88 *passim*
Griffiths, Gilbert, 140, 141, 171
Griffiths, Mrs. Samuel, 137
Griffiths, Samuel, 136, 138
Griffiths family, 136, 137-38, 139, 142, 144
Gunn, Giles, 47

Hapke, Laura: *Tales of the Working Girl,* 153
Harlan, County, Ky., 57
Harriman, E. H., 84
Harvard University, 52
Hawthorne, Nathaniel, 58
Heilbrun, Carolyn, 181
Henry, Arthur, 61
Hochman, Barbara, 57n44, 67n5
Hofstadter, Richard, 1, 81, 93n37
hooks, bell, 47
Horwitz, Howard, 65, 74, 120n19
Howells, William Dean, 23
Hume, David, 54-55, 58
Humma, John B., 165n23
Hurstwood, George, 64, 76-77n13, 86, 107, 117, 121-173 *passim*
Hurstwood, Jessica, 132
Hurstwood, Mrs., 129
Hussman, Lawrence E., Jr., 40n28, 179
Hutchisson, James M., 90n34

Identity, 4; and alienation, 177-78, 189; and business success, 80, 172-74; and class status, 118-19, 122-23, 130-47; and consumerism, 108, 109, 110, 112-13, 114, 115-28; and fashion, 125-26; and institutions, 136-37; reconstruction of, 135-37, 139-47; and sexuality, 168-70, 173-74, 176-77, 178; and work, 160-61
"Identity politics," 43
Idle curiosity, 17, 54, 90; defined, 49-50
Indiana University, 58
Industrial Workers of the World, 53
Industry: and machine process, 79-80; versus business, 18, 64, 66, 69-72, 77, 151
Instinct: development of, 164; role of, in behavior, 18-19, 152-62, 178, 188-89; role of, in criticism, 70-71. *See also* Parental bent; Workmanship
Institutionalism, 2, 3-4
Institutions: as "imbecile," 46; coercive control of, 175-76; impact of, 29; power of, 20-21; and predatory business principles, 52; versus revolutionaries, 84-85

INDEX   223

Intellectuals: as exiles, 50; Jewish, 56; need for rejection, 62-63; as outsiders, ix, x, 149; role of, 10, 11, 20, 21, 23, 46, 47, 50; and truth, 28

James, Henry, 131n38
James, William, x, 4, 13-14, 30, 32, 120n20, 159
Jameson, Fredric, 114, 143
Jephson, Reuben, 144
Jesus Christ: as businessman, 38
Johns Hopkins University, 52

Kane, Archibald, 178n34
Kane, Lester, 165n23, 166-85 *passim*
Kane, Robert, 17
Kansas City, 139, 140, 142, 144
Kaplan, Amy, 179
Kearney, Elizabeth, 10-11
Keller, Evelyn Fox, 35
Krafft-Ebing, Baron Richard von, 34
Kucharski, Judith, 180n38

Lasch, Christopher, 109, 141
Lawrence, Abbott, 52
Lears, T. J. Jackson, 112, 120n19, 120n20, 128
Lehan, Richard, 81
Lentricchia, Frank, x
Lerner, Max, 36n25, 63, 112
Lewis, Sinclair, 15, 62n51; *Elmer Gantry*, 38
Lieutenants of Finance, 69, 788, 86
Lingeman, Richard, 57
Literary realism. *See* Realism
Livingston, James, 114
Loeb, Jacques, 120n20
London, England, 99, 100
Lukacs, Georg, 145
Lycurgus, N.Y., 136, 137, 138, 139, 142, 144
Lynde, Polk, 78

McCord, Donald, 6
McCracken, Grant, 109
McDougall, William, 120n20
McTeague, Trina Sieppe, 166n25
Mailer, Norman, ix-x
Maine, Henry, 136n41
Marcus, Mordecai, 179

Marden, Orison Swett: *Success* magazine, 79
Martin, Ronald E., 81
Marx, Karl, 4, 55, 63, 71n7, 111, 160
Marxism, 12-13, 65
Masculinity, 41-42, 144, 167, 168, 169-70, 171-77 *passim*
Mason, Orville, 143, 144
Masters, Edgar Lee, 15
Meeber, Carrie, 5, 42, 64, 107-88 *passim*
Meeber, Minnie, 122
Mencken, H. L., 9, 12, 66, 26n9, 56, 57, 58, 61, 61-62, 88, 98, 122n24, 129, 151, 183, 189
Merton, Robert K., xn3, 4
Michaels, Walter Benn, 9-10, 65, 95n39, 114-15
Mill, John Stuart, 30n14
Mills, C. Wright, ix, x, 5-6, 48n34
Mitchell, Wesley, 2n2
Mizruchi, Susan, 6n8
Modernism, 15-16, 150; and postmodernism, 12, 13
Moers, Ellen, 80, 96, 130
Montreal, Canada, 132, 133
Morality: and business success, 171-73; in criticism, 36, 37, 83-85; and religion, 39; and revolutionaries, 83-85; and self-interest, 177-78; and superiority of women, 180. *See also* Business: tycoons as moral revolutionaries; Religion
Morgan, J. P., 112, 137
Motherhood: value of, 155-57, 165, 180. *See also* Feminism; Women's issues; Women's work
Myer, Gustavus, 45n101
Mystery Magazine, 137

Narcissism, 141
New Jersey, 111
New York, 59, 78, 126, 130, 131, 134, 137
New York Central Railroad, 158
New York Society for the Suppression of Vice, 56n43
New York Times Magazine, 108; "The Rich," 110-11
Nietzche, Fredrich, 63, 80

Nobel Prize, 62n51
Noble, David W., 82, 163-64
Norris, Frank, 22, 67; *The Pit*, 87
Novick, Peter, 35

Page, Walter, 61
Park, Robert, 5
Parental bent, 18, 154-89 *passim*. *See also* Instinct
Peirce, Charles Saunders, 12n20, 30
Petrey, Sandy, 114
Philadelphia, Penn., 91, 97, 99
Philadelphia "Big Three," 93
Pittsburg, Penn., 59
Pizer, Donald, 60, 88, 171
Poe, Edgar Allan, 58
Poirier, Richard, 181
Poor Richard, 99
Pope, Alexander, 29; *The Dunciad*, 38
Popular Science Monthly, 30
Porter, Noah, 82n23
Postmodernism, 31, 109n5
Pottery Barn (store), 108
Pragmatism, 12n20, 54; and truth, 30-31
Property ownership, 30n14
Psychoanalysis, 33, 34-35, 113
Puritanism, 9

Qualey, Carlton C., 7

Rabine, Leslie W., 124
Railroads: building of, 75, 77, 78, 96, 101, 102
Realism, 57; literary theories of, 150; and sentimentalism, 180-81, 184, 185
Religion, 21; challenge to, 43; and economics, 37-38, 39; and morality, 39; selling of, 45-46; and sexuality, 40, 41, 43-44; and skepticism, 93-94; teachings of, versus real life, 44, 85; versus science, 36, 40, 43
Riesman, David, 120n19, 155, 160
Riggio, Thomas P., 57n44, 114, 171
Rockefeller, John D., 52, 84, 87n29, 89, 100n44, 112, 113, 137
Rogers, H. H., 84
Rorty, Richard, x, 12, 16
Ross, Dorothy, 32n20, 48n34

Ross, Valerie, 175
Rubin, Gayle, 56n43

Sage, Russell, 84
Said, Edward, 10, 20, 36, 57; intellectual as exile, 49; *Representations of the Intellectual*, 47, 48
St. Louis, Mo., 59, 134
St. Louis Globe-Democrat, 146
Sartre, Jean-Paul, 10
Scharnhorst, Gary, 93n36
Schneider, Louis, 141
Scholarship: and academic freedom, 50; and blacklisting, 49-50; and self-interest, 38-39. *See also* Intellectuals
Schöpp, Joseph C., 66n3
Science: and skepticism, 54-55, 57; use of, in criticism, 31-32, 35-36, 37; versus religion, 36, 40, 43
Scott, Howard, 78
Sentimentalism: and business, 187; dialogue with, 186-87; in Dreiser's works, 164-65, 179-85; as feminine, 180, 181, 183, 186-87; as masculine, 187-88; reappraisal of, 181-82, 183-85; as second-rate, 179-80; as threatening, 184-85
Sexuality, 27, 28; American view of, 34; and class status, 168; and consumer culture, 118; in criticism, 43, 151, 155; and identity, 168-70, 173-74, 176-77, 178; as predatory, 169-70; and religion, 40, 41, 42, 43-44; and self-destruction, 129-30
Shannon, Christopher, 72, 153
Shloss, Carol, 83n25
Silverman, Kaja, 124
Simmel, Georg, 121
Simpson, O. J., 10
Sippens, Henry De Soto, 76
Slater, Gilbert, 113
Slough of Despond: psychology as, 113
Sluss, Mayor, 83
Smart Set, 61
Smith, Adam, 31, 37
Social class. *See* Class status
Social criticism. *See* Cultural criticism
Social Darwinism, 9, 27-28, 29, 81, 82, 83. *See also* Spencer, Herbert

Sollors, Werner, 46-47
Solomon-Godeau, Abigail, 118
Southworth, E. D. E. N., 186
Soviet Union, 86-87
Spencer, Herbert, 8, 27, 80, 81, 82, 83, 136n41, 164n21; *First Principles,* 93, 94
Standard Oil Company, 111, 112
Stane, Lord, 78, 79, 96
Stanford University, 52
Stener, George, 72, 88
Strychacz, Thomas, 143
Sumner, William Graham, 82
Sundquist, Eric, 3n5
Susman, Warren, 86n28
Swift, Jonathan, 38; "A Modest Proposal," 167

Tarbell, Ida M.: *The History of the Standard Oil Company,* 87n29
Tarde, G.: *Psychologie Economique,* 119
Tenure. *See* Universities
Thoreau, Henry David, 58
Tilman, Rick, 16, 148, 185
Tompkins, Jane, 180n38, 181, 183, 184
Trachtenberg, Alan, xn1, 114
Trilling, Lionel, 31n19
Trump, Donald, 79n16
Truth: in art, 25-27, 28; creation of, 46; critical evolution of, 22-23, 24, 29; and intellectuals, 28; and pragmatism, 30-31; redefining, 62; versus ceremonialism, 145-46
Twain, Mark, 25

Uncle Tom's Cabin, 184
U.S. Steel, 101
Universities: as businesses, 51-52, 58; and tenure, 50-52. *See also* Intellectuals; Scholarship
University of Chicago, 3, 52, 72, 100
University of Missouri, 15, 185

Valverde, Mariana, 125n30
Vanderbilt, William H., 84
Veblen, Andrew (brother), 7
Veblen, Ellen Rolfe (first wife), 15
Veblen, Thorstein
—Life: career as teacher, 3, 15, 16, 49-50, 185; emotional nature of, 185-86; family of, 7, 15, 16n28, 48; heroes of, 54-56; influence of, 7-8, 36n25; influences on, x, 4, 14, 16, 30, 32, 159; as outsider, 17, 21, 22, 46-48, 148-49; reputation as womanizer, 7; response to critics, 24; style of, ix-xii, 6, 12, 13-15, 16-17, 18, 21, 22, 29, 37, 38-39, 43, 49, 54-56; values of, 148, 158, 163-64; view on work, 158; works, renewed interest in, x, xi
—Works: *Absentee Ownership,* 15, 38, 106; "Christian Morals and the Competitive System," 38, 39; "Dementia Praecox," 33, 34, 35, 36; "The Economic Theory of Woman's Dress," 125; *The Engineers and the Price System,* 22, 52-53, 54, 78, 126, 187; *The Higher Learning in America,* 22, 49, 51-52, 53, 54; *Imperial Germany and the Industrial Revolution,* 14; *The Instinct of Workmanship and the State of Industrial Arts,* 148-89 *passim*; "The Intellectual Pre-Eminence of the Jews," 22, 55; *Laxdoela Saga,* 15; "Menial Servants during the Period of the War," 167; *On the Nature of Peace and the Terms of Its Perpetuation,* 63; *The Place of Science in Modern Civilization,* 22; "The Socialist Economics of Karl Marx and His Followers: I," 55; *The Theory of Business Enterprise,* 17, 65-106 *passim*; 106; *Theory of the Leisure Class,* xi, 1, 7, 12, 13-16, 18, 22, 24, 38, 48n34, 107-47 *passim*, 163, 167, 170, 174; *The Vested Interests and the Common Man,* 68
Vidich, Arthur J., 54n41

Ward, Lester Frank, 22, 163, 174-75
Warren, Robert Penn, 8
Warshaw, Jacob, 15, 185-86
Waterman and Company, 99
Waterman family, 103

Wealth: in America, 110-12; versus worth, 134-35
Weber, Max, 75-76, 78
West, Cornel, x
West, James L. W., III, 149n2
Wharton, Edith: *The Age of Innocence*, 176
White, Morton, 30n14
Whitman, Walt, 58
Wiebe, Robert, 86
Wilson, Christopher P., 60n48, 65
Wingate, Stephen, 97
Witla, Eugene, 5, 25-27, 28, 29, 66, 83n25, 84, 86, 156, 174, 188
Woman's novel, 181
Women: conduct of, 27, 28; as conspicuous consumers, 163, 169, 174; defining qualities of, 151-52; and moral superiority, 180; as "more essential," 174-75; as prizes, 168, 169-71, 174, 176; as productive workers, 163
Women's issues, 10, 18, 19. *See also* Feminism
Women's work, 153; historical devaluation of, 165, 166-67; value of, 177, 180
Woolf, Virginia: *A Room of One's Own*, 168
Work: value of, 152-62; versus labor, 158-62. *See also* Industry
Workmanship: instinct of, 18-19, 70, 152-53, 158, 159, 161-62, 164. *See also* Instinct
World War I, 35
Worth: versus value, 149, 150; versus wealth, 134-35

Yerkes, Charles Tyson, 72, 79n15, 84, 87n29, 88, 89, 96, 97n41, 104n47

Zanine, Louis, 40n28, 80
Zionism, 55-56
Zola, Emile, 33

PERMISSIONS

Grateful acknowledgment is made for permission to reprint materials from the following sources:

A portion of chapter 1 first appeared as "Thorstein Veblen and the Rhetoric of Authority" in *American Quarterly* 46:2 (June 1994). Reprinted by permission of Johns Hopkins University Press.

A shorter version of chapter 3 first appeared as "The Psychology of Desire: Veblen's 'Pecuniary Emulation' and 'Invidious Comparison' in *Sister Carrie* and *An American Tragedy*," in *Studies in American Fiction* 21:2 (autumn 1993). Reprinted by permission of the publisher.

A portion of chapter 4 first appeared as "Jennie through the Eyes of Thorstein Veblen," in *Dreiser's "Jennie Gerhardt": New Essays on the Restored Text,* edited by James L. W. West III, University of Pennsylvania Press, 1995. Reprinted by permission of the publisher.

Excerpts from letters in Joseph Dorfman's papers used with permission of Joseph Dorfman Papers, Rare Book and Manuscript Library, Columbia University.

Excerpts from *An American Tragedy* reprinted by permission of the Dreiser Trust.

Excerpts from an unpublished interview of Dreiser and "The Professional Intellectual and His Present Place," an unpublished essay by

Dreiser, used with permission of Special Collections, University of Pennsylvania.

Excerpt from letter by Becky Veblen Meyers used with permission of Thorstein B. Veblen Collection, Carleton College Archives, Carleton College.